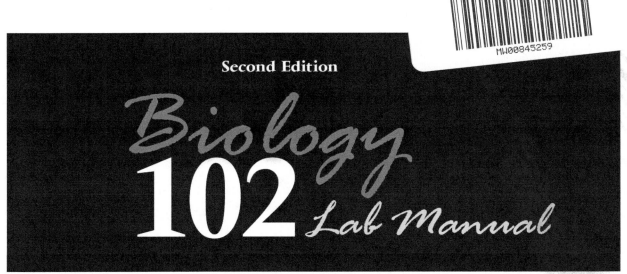

Second Edition

Biology
102 *Lab Manual*

William Gordon
Leon Dickson

Kendall Hunt
publishing company

Kendall Hunt
publishing company

www.kendallhunt.com
Send all inquiries to:
4050 Westmark Drive
Dubuque, IA 52004-1840

Copyright © 2011, 2016 by Kendall Hunt Publishing Company

ISBN 978-1-4652-6849-5

Contents

About the Authors

William R. Gordon earned his B.S. and M.S. degrees from Tuskegee University and his Ph.D. at the University of Minnesota, Twin Cities Campus. He received his Post Doctoral training at Brookhaven National Laboratory and Meharry Medical College. He served as a Program Director at the National Science Foundation and with his colleague, Dr. Kenneth L. Poff, he co-founded and conducted the Undergraduate Researchers in Plant Sciences Program at Michigan State University from 1993 until 2006. Dr. Gordon served as the Eastern Regional Vice-President for the Beta Kappa Chi National Scientific Honor Society for more than thirteen years. At Howard University, he taught nine different undergraduate and graduate courses in the biological sciences and served as a Coordinator of the Biology 101/102 introductory biology course for more than twelve years.

Leon A. Dickson, Jr. is an Associate Professor in Howard University ís Department of Biology. He earned his A.B. degree majoring in chemistry from Bowdoin College, his Ph.D. in biochemistry from M.I.T., and did his Post Doctoral studies in molecular biology at Purdue University. For ten years, Dr. Dickson was a faculty member in the Department of Biochemistry at Rutgers Medical School (now named Robert Wood Johnson Medical School) which was a component of the University of Medicine and Dentistry of New Jersey in New Brunswick, NJ. Dr. Dickson was a research associate at the Howard University Cancer Center and then joined the Department of Biology in August, 1990. His research has included the study of specific interactions of tRNAs and their aminoacyl-tRNA synthetases during protein synthesis, the exon-intron structure of chicken and human collagen genes, the basic molecular mechanisms of cancer, and the prediction of the tertiary structure of proteins from their DNA sequence (proteomics and genomics). With Professor Marilyn M. Irving (School of Education, Howard University), he has published on the blending of science content and pedagogy, retraining secondary science teachers, and designing effective professional development activities for improving science instruction by middle and high school teachers. Dr. Dickson has taught undergraduate and graduate molecular biology courses as well as taught in and coordinated General Biology.

Studying the Prokaryotes: Bacteria and Cyanobacteria

OBJECTIVES

At the conclusion of the exercise, you should...
1. recognize the differences between bacteria and other organisms studied so far.
2. know the different types of microscopic morphology of bacteria.
3. be able to recognize the shapes of bacteria.
4. be able to recognize the sizes of bacteria.
5. become familiar with observing bacteria that have been stained.

INTRODUCTION

The shapes of bacteria can be grouped into three types: **bacillus** (rectangular or rod-shaped), **coccus** (spherical or round), and **spirillum** (curved or helical). There are variations within each of these shapes. In this exercise, you will continue practicing using the microscope to examine the variety of shapes and sizes of bacteria.

MATERIALS

Supplies:

Prepared microscope slides of a gallery of bacterial shapes and sizes

PROCEDURES

Technical Background

Most bacteria range in size between 0.5-2.0 micrometers (μm). One aspect of bacterial identification is to describe their shape. The three basic types are rod, spherical, and curved. Rods can vary in length and width, may have square, round, or pointed ends, may or may not have flagella (for motility), and may occur singly or in chains. Rods that have variable sizes among individual cells (i.e., coccoid to long rods) are referred to as pleomorphic in shape. The spherical coccal cells may occur singly, in pairs, tetrads (groups of four), chains, or irregular clusters. The helical or curved bacteria can vary in length and width, and may occur as curved, bent, or wavy forms, with or without flagella. There are bacteria that form endospores, and this morphological characteristic is also used in classifying bacteria.

The following are examples of the bacterial shapes that will be observed in this laboratory exercise.

Gallery of Bacterial Shapes.

cocci: single, clusters, tetrads	Cocci: single, pairs, chains	cocci: diplococci	rods: large; single, chains, with endospores
1	2	3	4
rods: curved or spiral	rods: pleomorphic	rods: long, single, chains	rods: short (coccobacilli), single, chains
8	7	6	5

Observe the different shapes of bacteria in each section of the slide. Draw your observations on the next page. (Note: If gallery slides are not available, observe individual slides of the various shapes that are available.) Drawings are not to scale.

Examples of species of bacteria exhibiting the various shapes:

(**Note:** In most cases, the bacteria have been stained with a simple stain, unless otherwise noted.)

1. cocci–single and clusters: *Staphylococcus aureus*
   cocci–tetrads: *Micrococcus luteus*
2. cocci–single and chains: *Streptococcus* species
3. cocci–diplococci: *Neisseria* species (*Moraxella catarrhalis*)
4. rods–single and chains: *Bacillus megaterium* (some might show endospores,
   which do not stain, and will thus appear clear)
5. rods–coccobacilli: *Escherichia coli*
6. rods: *Pseudomonas aeruginosa*
7. rods–pleomorphic: *Corynebacterium* species
8. rods–curved or spiral: *Rhodospirillum* species or *Spirillum volutans*

Observe separate slides of the following:

1. spirochetes: *Treponema pallidum*
2. spiral: *Spirillum volutans*
3. mixture of cocci, rods, and spirals
4. rods with endospores
5. bacteria, yeast, and blood cells for size comparison

Examining Different Shapes of Bacteria

1. Use the Gallery of Bacteria Key and other slides to examine the different shapes of bacteria.
2. Make sure you feel confident in recognizing the different shapes, sizes, and arrangements of bacteria.
3. Draw your observations in the Evaluation of Results section.

EVALUATION OF RESULTS

Purpose

Data

Draw the organisms seen under oil immersion (100×) from the "Gallery" slide in the spaces below.

Gallery of Bacteria Key.

	cocci: single, clusters, tetrads	Cocci: single, pairs, chains	cocci: diplococci	rods: large; single, chains, with endospores
	1	2	3	4
	rods: curved or spiral	rods: pleomorphic	rods: long, single chains	rods: short (coccobacilli) single, chains
	8	7	6	5

Use the circles provided for drawing more of the organisms seen under oil immersion (100×) from the slides provided.

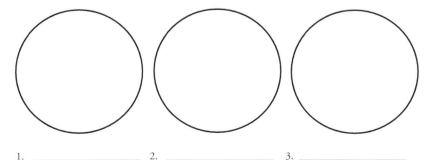

1. _____   2. _____   3. _____

Data continued:

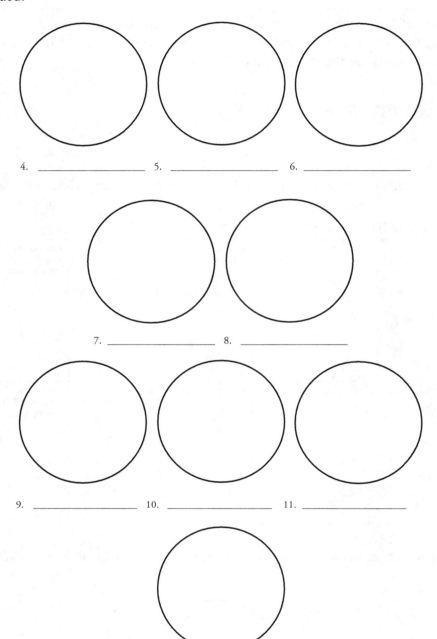

4. _____   5. _____   6. _____

7. _____   8. _____

9. _____   10. _____   11. _____

12. _____

Conclusions, Discussions, and Questions

1. Describe two of the differences between bacteria (procaryotes) and some of the eucaryotes studied so far.

2. List the three major shapes of bacteria, and give examples of each.

THE CYANOBACTERIA

Domain: Bacteria

Prokaryotic: Cells lacking a nucleus. Based on the sequences of ribosomal DNA, this group is only distantly related to the other domain of prokaryotes, the Archaea.

Cyanobacteria (blue-green algae)

The Cyanobacteria are photosynthetic bacteria utilizing both photosystems I & II, hence, envolving oxygen. This group is commonly referred to as the blue-green algae. Some members of the group fix nitrogen and are ecologically important both as primary producers and for their nitrogen fixation. Members process only one form of chlorophyll (Chlorophyll a). Accessory pigments include carotenoids, but the presence of the phycobilins (phycoerythrin and phycocyanin) are a distinctive feature of the group.

I. OSCILLATORIA

You are to know this genus.

PROCEDURE

At the front bench is a saucer of soil. The greenish-black film on the soil is mostly made up of the cyanobacterium *Oscillatoria*. There are many other interesting organisms, however, such as diatoms, euglenoids, and nematodes.

To prepare a slide, take a small piece of this film with a needle, place it in a drop of water on your slide, mix it around with your needles and cover with a coverslip. *Oscillatoria* is the filamentous organism that constitutes the matrix of the organic material on the soil.

Draw a filament of *Oscillatoria*. Label cells and filament.

II. ANABAENA

You are to know this genus, and to both recognize heterocysts and to know their function. You should also recognize the mutualistic relationship that exists between *Azolla* and *Anabaena*.

PROCEDURE

Anabaena will be found growing in the tissues of the water fern *Azolla* at the front. Take a small piece of tissue from *Azolla*, and it to a drop of water on a microscope slide, and tear it apart as finely as possible using teasing needles. Prepare a wet mount and observe with your microscope. Filaments of *Anabaena* should be clearly visible in the medium around the fern tissue.

Draw a filament. Be sure that you observe, draw and label at least one heterocyst.

The water fern *Azolla* forms a mutualistic relationship with a species of *Anabaena*. The plant provides a habitat for the cyanobacterium in pockets in its tissue. The plant, in turn, benefits from the nitrogen fixed by *Anabaena*. Because of this relationship, *Azolla* is a valuable *green manure* crop in flooded rice paddies in eastern Asia. While the fields are flooded, the plant forms dense mats over the water's surface. When drained, the plant and its symbiont decay, liberating nitrogen to the soil.

III. STROMATOLITES

Recognize these fossils as evidence of Cyanobacterial growth.

DEMONSTRATION

These layered sedimentary rocks are a byproduct of the metabolic activity of microorganisms, especially of filamentous cyanobacteria. Stromatolites are found in rock layers some of which are dated at over two billion years old. Until 600 million years ago the cyanobacteria dominated the landscapes of Earth, and were responsible for initiating the oxygen rich atmosphere that exists today. They formed extensive reefs composed of rocks such as this stromatolite. Stromatolites are still produced today, but only in environments which are too harsh for their predators to survive.

Topics for Discussion

1. How are the Cyanobacteria similar to the chloroplasts of plants? How are they dissimilar?

2. The heterocyst is the site of nitrogen fixation in the Cyanobacteria. This process requires anoxic conditions.

   a. From your text - does any type of photosynthesis occur in the heterocyst?

   b. How does the heterocyst survive if it can't fix carbon?

   c. How do the other cells in a filament of *Anabaena* benefit from the nitrogen fixed by the heterocysts.

3. Fossils of the same form as *Anabaena* and *Oscillatoria* have been discovered that are over 2 billion years old. Speculate about why certain forms may be retained while others are subject to rapid evolutionary change.

Studying Kingdom Protista: Protozoa

OBJECTIVES

At the conclusion of the exercise, you should ...

1. be able to prepare wet mounts of various living microorganisms.
2. observe microscopic features of protozoa, and algae.
3. recognize the diversity of microorganisms.
4. understand the basic differences between Procaryotes and Eucaryotes.
5. understand what hay infusion is.
6. be introduced to common genera and species of Eucaryotes.
7. know what a parasite is.

INTRODUCTION

The microbial world includes a wide variety of microorganisms. Samples taken from ponds and oceans will have many different species of protozoa, algae, plankton, and diatoms. This exercise will give you more practice using the microscope and introduce you to the diversity of the microbial world.

MATERIALS

Cultures:

Hay infusion

Euglena

Plankton tow

Pond water

Supplies:

Slides and coverslips

Lens paper

Pasteur pipettes with bulbs

Prepared slides of some medically important protozoa

PROCEDURES

Technical Background

Hay Infusion: An infusion is made by soaking dried plant material in water. For example, after soaking hay in non-chlorinated water for a week, a "hay infusion" will develop, yielding a variety of microorganisms. Protozoa are typically found in water, but when they become dry, they can form cysts and go into a dormant state. These cysts can be activated by adding water. At the beginning of the infusion, bacteria will predominate. Then a variety of other microorganisms will appear; these will include saprophytic flagellates, ciliates, and amoebae that feed on the bacteria. Eventually, the carnivorous ones will appear. The appearance of each new species is related to factors such as light intensity, gases present, pH, and concentration of organic compounds. The sequence of appearance of organisms in the hay infusion is representative of a food chain.

Succession: Succession refers to orderly sequential changes in the composition or structure of an ecological community. Succession in a protozoan community may be demonstrated in pond water.

Procaryote: A cell whose genetic material is not enclosed in a nuclear envelope and usually has a single circular DNA molecule as its chromosome. The bacteria are in this group.

Eucaryote: A cell having DNA inside a distinct membrane-enclosed nucleus (true nucleus) and usually other organelles. The protozoa and fungi (as well as animals and plants) are in this group.

Protozoa: Mostly unicellular eucaryotic microorganisms that lack cell walls.

Parasite: A type of organism that feeds on live organic matter, such as another organism.

Algae: The common name given to a heterogenous group of plants that are capable of carrying on photosynthesis and usually live in water. Some are unicellular.

Euglena: A one-celled organism that is usually green in color, can generally make its own food by photosynthesis (phototrophic), and is free moving.

Tetrahymena: A protozoan possessing cilia for motility.

Plankton: Free-floating aquatic microorganisms. This group includes diatoms and dinoflagellates. Diatoms have cell walls that consist of pectin and silica.

Making Wet Mount Slides

1. After viewing the videos showing some of the microorganisms that can be found in the environment, examine the various samples to try to find examples.
2. Place a drop of one of the liquid specimens on a microscope slide, cover with a coverslip, and examine with the brightfield microscope, starting with 10× and going up to 40×.
3. Use the manuals and diagrams provided, as a guide.
4. Note the means of motility (flagella, cilia, ameboid).
5. Note the difference in sizes of the bacteria and the protozoa. Protozoa can range in size from 1 mm to 70 mm or larger. Bacteria usually range from 0.5–2.0 mm.
6. Practice making wet mounts of all the samples, and record the observations in the Evaluation of Results.

Examples of Protozoa and Algae

Examples of Protozoa and Algae

Euglena

Tetrahymena

Paramecium

Amoeba

Algae

Examples of protozoan parasites

Giardia trophozoite

Trichomonas vaginalis

EVALUATION OF RESULTS

Purpose

Data

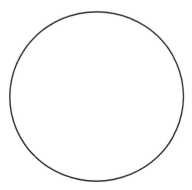

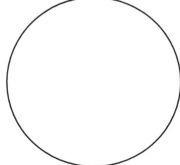

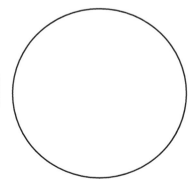

Specimen: _____

Total Mag.: _____

Specimen: _____

Total Mag.: _____

Specimen: _____

Total Mag.: _____

Data continued:

Specimen: _____

Total Mag.: _____

Specimen: _____

Total Mag.: _____

Specimen: _____

Total Mag.: _____

Specimen: _____

Total Mag.: _____

Specimen: _____

Total Mag.: _____

Specimen: _____

Total Mag.: _____

Conclusions, Discussions, and Questions

1. Explain two differences between protozoa and algae.

2. Use your textbooks and the Internet to complete the table below.

Table 2.1 Examples of Human Pathogens That Are Protozoa.

Name of pathogen	Disease	Source of human infections
Giardia lamblia		
Entamoeba histolytica		
Naegleria fowleri		
Cryptosporidium species		
Trichomonas vaginalis		

3. Write a short paragraph in which you compare what you can see in the pond water or plankton tow with your naked eyes with what you can see through the microscope at 100× and 400× total magnification.

Studying Kingdom Protista: Algae and Slime Molds

DOMAIN: EUKARYA - ORGANISMS WITH NUCLEATED CELLS
KINGDOM PROTISTA - THE PROTISTS

The protists are a loose assemblage of organisms. Certain members have stronger ties to the plants, or to the animals or to the fungi than to other members of the group. It is largely a catch-all of eukaryotic organisms that don't properly belong in some other Kingdom. They are characterized by having a low level of cellular differentiation and, while some are multicellular, in these cases their bodies are relatively simple and are usually referred to as a **thallus**.

Phyla considered in this topic:

  I.   Division Chlorophyta - Green Algae
 II.   Division Bacillariophyta - The Diatoms
III.   Division Phaeophyta - The Brown Algae
IV.   Division Rhodophyta - The Red Algae
 V.   Division Oomycota - The Water Molds
VI.   Division Myxomycota - the Plasmodial Slime Molds

I. DIVISION CHLOROPHYTA - GREEN ALGAE
THE CHLOROPHYTA

Of all the divisions in the kingdom Protista, this is the one which is most directly relevant to botanists. This is because plants are believed to have evolved from the green algae. There are several lines of evidence that lead us to this conclusion:

1. Green algae have the same photosynthetic pigments as plants (Chlorophylls a & b, xanthophylls and carotenoids).
2. They store their food as starch in plastids (chloroplasts) - unlike plants (except hornworts!), starch deposition is associated with a structure called a **pyrenoid**.
3. Some have cell walls composed of cellulose.
4. One class of the Chlorophyta, the Charaphycean green algae, have members that undergo cytokinesis, like plants, through the creation of a cell plate mediated by a phragmoplast.

Today you will see seven genera of Chlorophyta. Observe them carefully and sketch each one. In each case, note the grass-green coloration which is a reflection of their pigmentation. Identify chloroplasts in each case and pyrenoids if apparent. You need to recognize each to genus, class, phylum and kingdom.

Domain: Eukarya

Kingdom: Protista

Division: Chlorophyta

    I.   Class: Chlorophyceae

   II.   Class: Ulvophyceae

  III.   Class: Charophyceae

I. Class Chlorophyceae

Chlamydomonas: Prepare a wet mount of this motile, unicellular green alga. While it requires some effort, look for the two flagella at the anterior end and the red eye spot. Look for cells that are moving but seem stuck in one place. In the center of the chloroplast a pyrenoid should be apparent. Compare your view with that of the electron micrograph by the culture.

Pandorina: This alga consists of colonies of Chlamydomonas-like cells arranged in a sphere with their flagella projected outward. Observe the hanging-drop slide of free swimming colonies. Now prepare a wet mount of your own, but first observe without a cover slip. Then add a cover slip and observe at 400× to clearly see details of the individual cells. Draw both the colonies and the cells.

Volvox: This impressive alga again consists of colonies of Chlamydomonas-like cells. In this case, however, each colony includes over 500 cells. As before, observe the hanging-drop slide of free swimming colonies. Now prepare a wet mount of your own, but first observe without a coverslip. Then add a cover slip and observe at 400× to clearly see details of the individual cells. Draw both the colonies and the cells.

These three genera are examples of the Volvocine line. The colonial forms must have evolved from a *Chlamydomonas*-like ancestor.

Can you speculate as to the selective pressure driving this evolutionary trend towards multicellularity?

II. Class Ulvophyceae

Ulva: Observe the example of "sea lettuce". Prepare a wet mount of the tissue. Identify the chloroplasts and pyrenoids. By through-focusing verify that this thallus consists of two distinct layers. Draw both the whole organism and a view of its cells.

III. Class Charophyceae

Spirogyra: Prepare a wet mount of this filamentous organism. Note that the filaments are unbranched. The cells have spiralled chloroplasts with obvious pyrenoids. Using through-focusing, search for nuclei in the center of each cell. Like plant cells viewed earlier, most of the cell's volume is occupied by a huge central vacuole. The nucleus is in an island of cytoplasm suspended inside the vacuole.

Desmids - *Cosmarium* (know common name): These are unicellular, though their bilateral symmetry often makes them appear as two-celled. Prepare a wet mount. Note chloroplasts and pyrenoids. By through-focusing you may be able to discern Brownian motion of gypsum particles in their vacuoles.

Chara: This is a morphologically complex alga. Take a Petri dish with a sample to your seat. Make a quick sketch. Make a wet mount of a fine filament at the end of one of its whorled branches. **Note** the disk-shaped chloroplasts. They are similar to those observed in *Elodea*. Like plants, *Chara* and *Coleochaete* (see below) both produce a phragmoplast during cell division and both belong to the clade of green algae from which the plants evolved. *Chara* also has sperm which closely resembles those of mosses.

Coleochaete: This is also morphologically complex alga. Make a wet mount of one of the disc-shaped colones. These cells make up a parenchyma-like tissue. Like in plant tissues, plasmodesmata interconnected adjacent cells. **Note** the large chloroplast in each cell. Draw both a view of the whole colony and another illustrating cellular detail. In the second drawing label both the chloroplast and its associated pyrenoid.

II. DIVISION BACILLARIOPHYTA - THE DIATOMS

This is one of three phyla of heterokont protists. The diatoms are photosynthetic and possess chloropylls a and c. Cell walls are made of silicon. The group is among the most important primary producers of the world's freshwater and marine environments. While diatom cells can remain attached after cell division to form simple colonies, the group is entirely unicellular. The cell walls consists of two parts called valves that are constructed somewhat like the two parts of a petri plate, with the top part overlapping the bottom. During cell division one daughter cell receives the bottom valve and the other the top. A new bottom then develops in each. Because of the structure of the cell walls, diatoms have two distinctive profiles. The **valve** view is the view seen when we look face on at one of the valves. The **girdle** view is the profile seen when the junction of the two valves faces the viewer. Based on the symmetry of the valves, two groups of diatoms are recognized. The pennate diatoms which have valves that are bilaterally symmetrical, and the centric diatoms which have valves that are radially symmetrical. Note that in each case the girdle views are rectangular.

Centric Diatoms:Valve view on top, bottom shows eight cells in a filament all in girdle view.

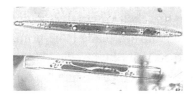

Pennate Diatom:Valve view on top, bottom shows girdle view of the same cell.

Objectives:

Recognize diatoms to phylum. Also be able to identify diatoms as pennate or centric, and know the difference between a valve view and a girdle view of a cell.

IIa. Centric vs. Pennate Diatoms: Observe the demonstration slide of pennate and centric diatoms. These are displayed in the *valve* view.

IIb. Freshly collected diatoms: Sample material from the bottom of the bowl of the freshly collected diatoms and make a wet mount.

IIc. Diatomaceous Earth. Take a sample from the container of diatomaceous earth and make a wet mount: add a drop of water to the slide, wet a teasing needle in the water, put the moist needle tip into the diatomaceous earth, and then stir the coated tip in the drop of water and add a cover slip.

Drawings: Draw examples (at least two) of diatoms both from the freshly collected material, and/or from the diatomaceous earth. Indicate if the diatoms observed are centric or pennate, and if your drawing represents a valve or girdle view (ideally show both from each example).

III. DIVISION PHAEOPHYTA - THE BROWN ALGAE

This is one of three phyla of heterokont protists. The common name "brown" algae is derived from their color, which is due to the accessory pigment fucoxanthin. Like the diatoms their plastids contain chlorophylls a and c. The group is entirely marine and are multicellular "sea weeds." Algin, a commercially important emulsifier, is produced from the kelps. Of all the protists, members of this phylum have achieved the highest degree of morphological and cellular differentiation. Several different groups have developed a degree of structural complexity such that their external structure (their morphology) can be categorized into three parts: holdfast (anchoring the organism), blade (the primary photosynthetic structure) and stipe (the part connecting the blade to the holdfast). Further, in the kelps, there occurs a high degree of cellular differentiation. Some kelps have cells that function and look like the cells in the phloem of higher plants that serve to move photosynthate through the organism.

Objectives:

Recognize examples to phylum. Also be able to identify the branching pattern of *Fucus*. If applicable, recognize holdfast, stipe and blades of examples seen in lab. Recognize sieve-tube members in the kelps and know why they are adaptive to the survival of these organisms.

IIIa. Fucus: Rockweed

Carefully observe the living material at the front of each bench. Note the holdfast. Now examine the opposite extremities of the organism. Growth in *Fucus* is like that in plants. It occurs at an apical meristem.

See if you can identify a region of cell division (an apical meristem).

Now observe the pattern of branching.

Based on your observations how does branching arise in *Fucus*?

While *Fucus* is not related to plants, its external morphology is very much like two plants you will study next week, *Marchantia* and *Psilotum*. These similarities are due to convergent evolution (The development of similar structures by unrelated organisms due to similar environmental pressures). The similarities between *Fucus* and these primitive plants are like the similarities between the wings of birds and those of insects.

Draw *Fucus*: label **holdfast** and clearly illustrate the pattern of branching

IIIb. Sargassum

If you frequent the beaches of the east coast, this is the one brown alga you are most likely to encounter. It is morphologically complex. It has obvious parts which can be described as blade and other parts which can be described as stipe, as well as having air bladders. Some *Sargassum* is found in floating masses at sea (The Sargasso sea gets its name from Sargassum). Other material from the same species are anchored by holdfasts. Interestingly, only the material that is anchored reproduces sexually. The masses adrift at sea reproduce vegetatively only by fragmentation. These floating masses support a rich community of organisms in areas that would otherwise be sterile.

IIIc. Kelps

The most complex brown algae are the kelps. All the kelps have body parts that can be divided into **stipe, blade, and holdfast**. Some members of the group also have specialized cells to conduct photosynthate. Since some kelps can grow hundreds of meters long, it is not surprising that they evolved cells that allow photosynthate to move from the illuminated blades to the shaded stipes and holdfast.

Observe the examples of preserved kelps on the side bench.

Draw a kelp: label **blade, stipe, and holdfast.**

Observe the demonstration of a sieve-tube member in *Macrocystis*. Sievetube members of the higher plants move photosynthate in the form of sucrose. Sieve-tube members of the kelps move mannitol.

Take some time on your own to observe the various examples of kelps illustrated through our web page.

IV. DIVISION RHODOPHYTA

This is primarily a marine group of photosynthetic organisms. Members lack flagella, their pigmentation includes chlorophyll a and phycobilins (phycoerythrin and phycobilin). These are the same type of pigments as is found in the Cyanobacteria, and it is likely that their chloroplasts were derived from an endosymbiotic cyanobacterium. The common name, red algae, is derived from their color, which is due to the accessory pigment phycoerythrin. The group is largely made up of multicellular "sea weeds" which have an alternation of generations (each species have free-living haploid and diploid vegetative stages). Some members (the coralline red algae) are important reef builders. Agar is produced from the red alga, Irish Moss. *Porphyra,* is a commercially grown red alga, and is processed into Nori, a staple in the Japanese diet

Objectives:

Recognize examples seen in lab to phylum. Know why their color is adaptive for life in deep water.

Observe the examples of red algae on demonstration.

Some members of the Rhodophyta live deeper in the ocean than any other photosynthetic organism. How are their accessory pigmentations adaptive for these conditions?

V. DIVISION OOMYCOTA - THE WATER MOLDS

This is the first of three heterokont phyla of protists we will study in lab. Studies of ribosomal RNA and structural similarities of the flagella between these groups indicate a common ancestry. Of the three phyla we study in lab this is the only heterotrophic group. The oomycota are similar to the fungi and, in the past, have been considered fungi. They are heterotrophic, and their bodies are made up of hyphae that are collectively called a mycelium. These hyphae are coenocytic. Unlike the fungi, they are a diploid group that produce gametes by meiosis, their cell walls are made of cellulose (not chitin), and their flagellated cells have two flagella, one of which is a tinsel flagellum. This last character firmly associates the group with other phyla of the protists and not with the fungi. The oomycota include members responsible for several important plant diseases.

Objectives:

Know the genus *Saprolegnia* and recognize the following: coenocytic hyphae, zoosporangia, oogonia, antheridia, zygotes.

PROCEDURE

Saprolegnia growing on hemp seeds are in petri dishes at the front. *Saprolegnia* is a saprophyte that can turn parasitic (see the interesting article on mouth fungus at the front). The petri plates are numbered: the oldest dishes are labelled with the smaller numbers. Younger cultures have better zoosporangia, the older ones have better gametangia.

1. **Zoosporangia** Take a dish with a larger number. With the 4x objective in place, and with the cover of the petri plate off, carefully place the dish on the mechanical stage of your microscope. Look for hyphae with dense cytoplasm in the new growth at the periphery of the mycelium. After locating hyphae with dense cytoplasm, carefully switch to the 10x objective and look for a septation cutting off the area of dense cytoplasm from the coenocytic mycelium. Typically zoosporangia are sausage-shaped. Once identified look for evidence of maturing zoospores in the zoosporangium. If you can identify maturing zoospores, share your slide with your TA.

   Observe the figures on the next page and use them for reference while viewing your culture.

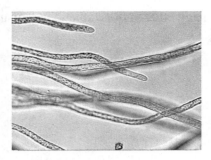

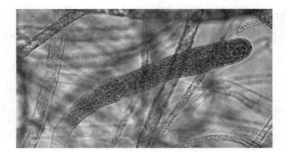

Coenocytic Hyphae   Zoosporangium

2. **Gametangia** *Saprolegnia* is homothallic. Further the gametes are never flagellated. Spherical hyphae called oogonia produce eggs by meiosis. Other hyphae from the same mycelium will grow through the oogonium and will undergo plasmogamy with the eggs to deliver sperm nuclei. These hyphae are antheridia. Observe a number of different cultures of varying ages. The oldest cultures will have oogonia with dense zygotes. Antheridia will still be visible attached to these oogonia. Younger cultures should have translucent oogonia.

Label the figure of younger gametangia to the right.

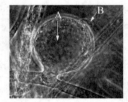

A. _____

B. _____

Draw an oogonium with zygotes from an older culture.

VI. DIVISION MYXOMYCOTA

The plasmodial slime molds. This is a prime example of a protist that did not fit into the old two kingdom system. It has a duality that was both animal-like and plant-like. The plasmodial stage consists of a coenocytic ameboid mass. This plasmodium engulfs prey by phagocytosis and responds to external stimuli. When food or water are scarce, however, the entire mass transforms itself into fungal-like sporangia.

Objectives:

Recognize plasmodia and sporangia of plasmodial slime molds, and cytoplasmic streaming in the plasmodium.

VIa. The Plasmodium

Take a water agar culture of *Physarum* from the front bench. Note the pattern presented by the plasmodium on the agar. Remove the cover of the petri plate and, with the 4x objective in place, position the entire plate onto your mechanical stage. Bring the plasmodium into focus and observe the flow of the food vacuoles due to cytoplasmic streaming.

Is the flow always in one direction?

If you have a second hand on your watch, time the flow from one reversal to another. Time at least three intervals. What are the times? Do they vary?

Compare your results with others in the class and record these comparisons below:

Speculate about how the oscillation of the flow of the cytoplasm may be adaptive.

Draw a plasmodium as it appears to the unaided eye on the agar.

VIb. Sporangia

View the examples of plasmodial slime mold sporangia on demonstration.

Next to your drawing of a plasmodium of *Physarum*, draw its sporangium.

Studying Kingdom Fungi: Molds, Sac Fungi, Mushrooms and Lichens

DOMAIN: EUKARYOTA - ORGANISMS WITH NUCLEATED CELLS
KINGDOM: FUNGI

Historically fungi have been considered to be plants. Molecular evidence, however, indicate that they are actually more closely allied with the animals. The group is distinctive enough that it is assigned the rank of a kingdom with four phyla.

Fungi are all **heterotrophic**, and survive either as **saprophytes**, **parasites**, or in **mutualistic partnerships**. Fungi have **cell walls made of chitin**, and with the exceptions of some unicellular species, have bodies composed of filaments called **hyphae**. Masses of hyphae are called a **mycelium**. One phylum, the Chytridomycota, have alternation of generations, have flagella and are aquatic. The inclusion of the Chitrids in the fungi is a recent determination based on molecular evidence. The other three phyla all lack flagella, are primarily terrestrial and lack an alternation of generations (they all have zygotic meiosis). Membership in these three groups is primarily based on the spore bearing structure developed after meiosis. The Zygomycota have zygosporangia, the Ascomycota have asci, the Basidiomycota have basidia. In today's lab we will study examples from these last three phyla.

I. DIVISION ZYGOMYCOTA

Members with bodies made up of **coenocytic** hyphae and that produce **zygosporangia**. The group is important as saprophytic decomposers and as mycorrhizal symbionts.

Objectives:

You should recognize the genus *Rhizopus*, and the following types of hyphae from that genus: stolon, rhizoids, sporangium, sporangiophore. You should recognize gametangia and zygosporangia from either *Rhizopus* or from *Phycomyces*, and recognize that the presence of this structure defines this phylum. You should also recognize coenocytic hyphae.

Ia. Sexual Cultures of Phycomyces on the Side Bench

Take one of these cultures to your seat. Note the two points of inoculation on either side of the plate. *Phycomyces* is heterothallic - that is it has genetically determined mating types which must out-cross with strains of a different type. In this case there are two strains, "+" and "−". Sexuality is manifested by a line running midway between where each strain was inoculated. Remove the cover and observe this region of sexuality more closely using a dissecting microscope.

Draw a pair of gametangia.

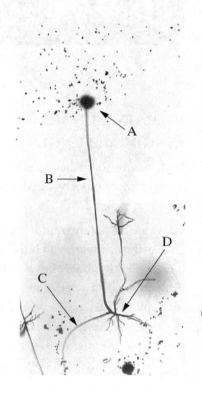

Ib. Genus Rhizopus

Take a culture of *Rhizopus* growing in a mini petri dish located on the side bench to your seat. Do not remove the cover! Observe using your compound microscope with the 4× objective. Identify the various types of hyphae illustrated to the right.

Note that the organism is coenocytic except where the spores are formed in the sporangium.

A = _____

B = _____

C = _____

D = _____

Take a prepared slide of *Rhizopus* and look for zygosporangia and gametangia.

Draw a zygosporangium.

II. DIVISION ASCOMYCOTA

Hyphae with incomplete septa. Plasmogamy and karyogamy separated in time, resulting in a persistent dikaryotic stage. Members form **asci**, which are often associated with a fleshy fruiting body. Many members reproduce asexually by conidia borne on conidiophores. The group is important as decomposers, parasites, and symbionts. They are the primary group of fungi that form lichens and, while not as widespread as the Zygomycota, are important in mycorrhizal associations.

Objectives:

Recognize and be able to name the following: asci and ascospores; budding cells of yeast; fruiting bodies of *Sordaria*, *Morchella* and powdery mildew; conidia of *Penicillium,* know where spores are borne on a morel.

Know the following groups

>   Genera: *Morchella, Penicillium*
>   Common Names: Powdery Mildew, yeast

IIa. Genus Sordaria

Cultures on the side bench. **Sordaria** is homothallic. Any genetic type can mate with itself as we observed in the meiosis activity.

Procedure:

Crush a fruiting body as outlined in the meiosis lab and observe the asci.

Draw an ascus

IIb. Genus Penicillium

Cultures on the side bench. The blue coloration is due to a type of asexual spore borne by various members of the Ascomycota called **conidia**. Members of this genus are the source of penicillin, the first antibiotic to be identified and used in medicine.

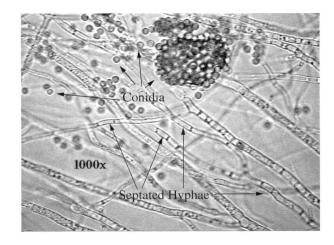

Procedure:

Prepare a wet mount of the material at the very surface. Choose an area that is blue. Identify the structures illustrated. Also observe the conidia on the intact conidiophores through the demonstration microscope.

IIc. Yeast

Budding culture at the front.

The type of yeast used in baking and brewing is a member of the Ascomycota. Yeasts are unusual in that they are unicellular. Yeast cells divide unequally and this pattern of division is termed budding.

Procedure:

Prepare a wet mount using a drop of the culture:

Draw budding yeast cells.

IId. Powdery Mildew - a Parasitic Ascomycete
Procedure:

Observe the infected leaves of nannyberry available at the front of each student bench. Note the white areas on the leaf that give the parasite its common name. These are areas where hyphae have broken through the leaf and bear conidia.

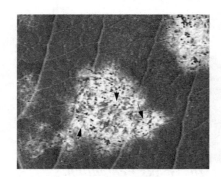

Infected leaf of Nannyberry with ascomata labeled

See the demonstration microscope of conidia of a powdery mildew on the center student bench.

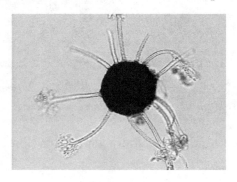

Fruiting body of powdery mildew

Now identify the black dots found within the white areas. These are fruiting bodies that totally enclose the asci. Typically these remain intact through the winter. Flood an area containing these ascoma with 70% alcohol and scrape a number off onto a microscope slide into a drop of water. Observe at 40×. See illustration to the right. Now, while observing through your 4× objective (your shortest lens), crush the fruiting body using a teasing needle by applying pressure directly over it. Apply only enough pressure to crack the outer surface! Now the asci inside will imbibe water, will swell and become visible within a minute or two. If you experience difficulties ask your TA for help.

Draw a fruiting body with emerging asci.

Draw an ascus with ascospores.

IIe. Genus Morchella

Preserved fruiting bodies on each bench and a prepared slide.

A morel is an example of a fleshy ascomycete. The fruiting body consists of three types of hyphae, two are from different mating strains, the third is the product of plasmogamy between these strains and is dikaryotic. While observing a morel, take the prepared slide through the fruiting body and determine where on the structure this section was cut. Observe the prepared slide with your microscope and find the red stained ascospores residing in asci.

Draw an ascus with ascospores.

Label where the asci are borne on the fruiting body pictured to the right.

III. DIVISION BASIDIOMYCOTA

Hyphae with incomplete septa; members produce basidiospores borne on basidia which are often associated with a fruiting body; plasmogamy separated in time from karyogamy resulting in a persistent dikaryotic stage.

Objectives:

Recognize basidia and basidiospores, and the cap, stipe and gills of a mushroom. Recognize all the examples of fruiting bodies seen in lab to phylum. Know where basidiospores are borne on a mushroom.

IIIa. Diversity of the Fleshy Basidiomycota

With a few notable exceptions, the fleshy fungi we typically observe outside are basidiomycetes. Mushrooms, jelly fungi, tooth fungi, shelf fungi, stinkhorns and puffballs are all basidiomycetes (the structures we observe are all fruiting bodies). See the display of various examples in your lab room.

IIIb. Observations of a Mushroom

Fresh mushrooms and prepared slide (section through a cap of *Coprinus*):

Take a mushroom and observe its gross morphology. Identify the stalk (stipe), and cap with the gills on the underside. Take the prepared slide of the section through the cap of *Coprinus* and, while using your mushroom as a reference, determine where the section was made. Now observe your prepared slide with your microscope and locate where the basidia are borne. Indicate on the illustration to the right where the basidia are borne.

Draw a basidium with basidiospores. Label basidium, basidiospores and sterigma.

IIIc. Other Basidiomycota

Rusts and smuts are parasitic members of the Basidiomycota. Observe the examples of wheat rust and corn smut in your lab room.

IV. LICHENS - A DUAL ORGANISM

Lichens consist of a fungus (almost always an ascomycete) associated with either a cyanobacterium or a green alga (a eukaryote with chloroplasts).

> **Activity:** Observe the lichens attached to the pieces of bark on the side bench. Break off a small piece using your forceps, place the material in a drop of water on a microscope slide and tease the material apart using your needles. Add a cover slip and observe with your microscope. Look closely at the smaller fragments to view the phycobiont associated with the fungus.

Is this organism a green alga (eukaryotic) or a cyanobacterium (prokaryote)?

Draw a cell of this photosynthetic organism. Label any fungal hyphae still attached.

Studying Plant Kingdom: Liverworts, Mosses, and Hornworts

DOMAIN EUKARYA
KINGDOM PLANTAE

Eukaryotic, photosynthetic organisms with chlorophylls a and b, xanthophylls and carotenoids, cell walls with cellulose, and with food stored as starch localized in plastids. As is found in the class Charophyceae of the Chlorophyta, cytokinesis is accomplished by means of a phragmoplast. Unlike the Chlorophyta, a higher level of cellular differentiation is found.

The structural complexity found in the plants is due to the environmental pressures associated with life on land. Starting with the primitive plants, a sterile jacket of cells surrounds the sexual structures preventing dehydration of the sperm and eggs (archegonia and antheridia though higher plants may lack one or both of these). Higher plants have tissues specialized for the transport of water and photosynthate, allowing plants to tap underground water and mineral resources and to support those tissues shaded from the light.

All plants are oogamous and have a heteromorphic alternation of generations with sporic meiosis, and gametic mitosis. All plants have embryos. The embryo is an early sporophytic (diploid) stage that is nourished by the gametophytic generation. This characteristic is the reason why one alternate name for the group was Embryophta.

THE BRYOPHYTES

The bryophytes are not a natural group. The three phyla are no more related to each other than they are to non-bryophytes. Historically, they have all been considered to be non-vascular plants. New studies, however, indicate that the mosses (Phylum Bryophyta) have cells specialized for the movement of water and photosynthate. There is also speculation that these cells are homologous to the vascular tissues of higher plants! All three phyla are unique in that the dominant generation is the gametophyte. In every case the sporophyte is dependent on the gametophyte for survival.

I. DIVISION HEPATOPHYTA: THE LIVERWORTS

One example: *Marchantia*

This is the most primitive phylum of plants. The group entirely lacks vascular tissues and stomata (there are air pores, but these are not associated with guard cells). The gametophytic generation is the dominant generation being free-living and photosynthetic. The sporophytes are totally

33

dependent on the gametophyte for survival. While the tissues of the gametophyte are undifferentiated and form a *thallus,* the cells are interconnected by plasmodesmata as are the tissues of all plants.

Objectives:

Recognize *Marchantia* to phylum and genus, and its **gametophytic** and **sporophytic** generations. From the gametophytic generation recognize the **thallus**, **air pores** in the thallus; **antheridiophores**, **antheridia** in the antheridiophore; **archegoniophores** and **archegonia** in the archegoniophores. Recognize the **spermatogenous tissue** and **sterile jacket** of the antheridium. Know that **sperm nuclei are produced by mitosis**. Recognize the following parts of the archegonium: **egg, venter, neck.** Know that **egg nuclei are produced by mitosis**. In the sporophyte recognize the **foot, calyptra, sporangium**, and **spores**. **Know that spore nuclei are produced by meiosis.**

Ia. The Gametophyte

1. The Vegetative Thallus

Observe the *Marchantia* culture at the front of the room, and take a petri dish of *Marchantia* to your seat.

The non-reproductive portion of the plant grows firmly attached to its substrate.

How does the thallus branch as it grows?

Using a dissecting scope observe the surface of the thallus. Note the air pores that appear as dots on the top surface.

Now get the prepared slide of a cross section of the thallus and observe with your microscope.

Note the relatively undifferentiated nature of this plant body. There are no vascular tissues of any kind. Also note that only cells at the top portion of the thallus are photosynthetic.

How is photosynthate transported in this plant body?

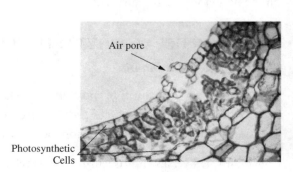

Look closely at a cross section of an air pore and note the absence of guard cells.

2. Sexual Structures

Observe the elevated umbrella-like structures growing up from the thallus, and note that there are two types. One has spokes radiating from the stalk like the ribs of an umbrella, the other has a stalk terminating in a disc. The structure with the spokes is an **archegoniophore and bears archegonia**. The structure with the disc is an **antheridiophore and bears antheridia**. In each case these structures are made up of gametophytic tissue.

Archegoniophores   Antheridiophores

Archegoniophores

Take the prepared slide labeled "*Marchantia* archegonium" and compare it with the live material in the petri dish. Place it on your microscope and locate the archegonia.

Draw an archegonium: label egg, venter and neck. Indicate above where the archegonia are located on the archegoniophore.

Antheridiophores

Take an antheridiophore and rub the top of its disc in a drop of water on a microscope slide. Make a wet mount and look for swimming sperm. If you locate any tell your TA and share this slide with your neighbors.

Take a prepared slide labelled "*Marchantia* Antheridia" and compare it with the antheridiophore in your petri dish, then place it on your microscope. Locate the antheridia embedded in the disk of the antheridiophore.

Draw a antheridium: label spermatogenous tissue and sterile jacket layer of cells.

Specifically how is the structure of the antheridiophore adaptive to *Marchantia*?

Specifically how is the structure of the archegoniophore adaptive to *Marchantia*?

What type of nuclear division results in the production of the sperm and egg nuclei in *Marchantia*?

3. The Sporophyte

The sporophyte develops while surrounded and nourished by the tissues of the gametophyte. This is a fundamental characteristic of all plants. The mature sporophyte of Marchantia can be found hanging from the archegoniophore.

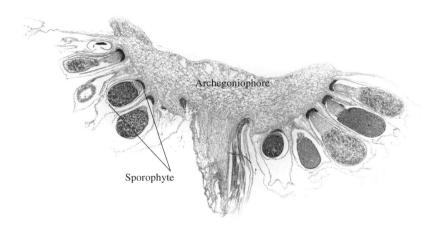

Take a petri dish with preserved sporophytes to your seat and observe the underside of the archegoniophores with a dissecting microscope. Note the bumps representing the mature sporophytes.

Now take the prepared slide labelled "*Marchantia*: Mature Sporophyte", and observe using your microscope. Identify the **foot, stalk** (seta), **sporangium**, and the **spores** in the sporangium. Label the figure below:

A = _____

B = _____

C = _____

D = _____

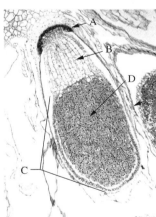

Note the unlabelled arrow indicating the calyptra. This tissue is part of the venter that enclosed the developing embryo.

What nuclear division results in the formation of the spore nuclei?

How might the elevation provided by the archegoniophore be adaptive to the sporophyte?

II. DIVISION ANTHOCEROPHYTA: THE HORNWORTS

This group is both more and less primitive than the liverworts. Like members of the Hepatophyta, there is little tissue differentiation in the gametophyte. The thin thallus grows against the substrate. The cellular structure also seems primitive. Its chloroplasts are algal-like. Most plant cells are like those we observed in *Elodea* and have numerous, disk-shaped chloroplasts . While starch is stored in plant plastids, it is usually not associated with pyrenoids like those of the green algae. Hornworts cells have one central algal-like chloroplast with a pyrenoid. This marks the group as being primitive. However, the hornworts have another character that is more advanced than the liverworts. The sporophytes of the hornworts have guard cells associated with the openings in its surface layer of tissues. This marks these openings as being true stomata - something not found in the liverworts.

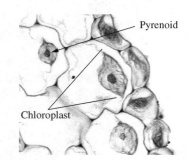

Cells of *Anthoceros* Gametophyte

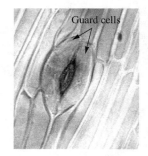

Stoma of *Anthoceros* Sporophyte

Consider the following evolutionary schemes:

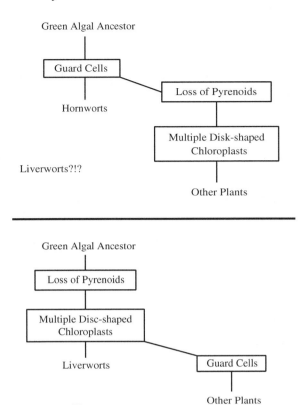

Which, if either, is "true" ?!?

Objectives:

Recognize hornworts to phylum. Know which part is gametophyte and which is sporophyte.

Observe the living material on the side bench. The **gametophyte** is a flat sheet-like thallus. The **sporophytes** are the "horns" growing on the gametophyte. Label the figure below:

A = _____

B = _____

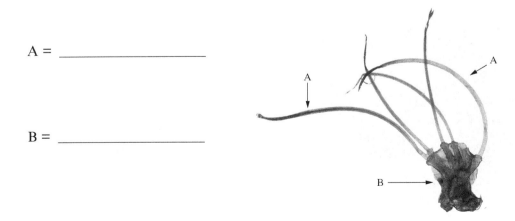

III. DIVISION BRYOPHYTA: THE MOSSES

Of all the bryophytes the mosses have the clearest affinity to the vascular plants. Within the Bryophyta, the true mosses have specialized cells that conduct water and others that conduct photosynthate. Further, these cells appear to be homologous to similar tissues in the vascular plants. Unlike the vascular plants, in the Bryophyta the gametophytic generation is the dominate generation, the sporophyte being dependent on the gametophyte for survival. The mosses also have disc shaped chloroplasts, lack pyrenoids and have stomata.

Objectives:

Recognize these plants to phylum, and able to differentiate gametophytes from sporophytes. From the gametophytic generation. Recognize antheridia and archegonia and the **calyptra associated with the sporangium.** Recognize the **spermatogenous tissue** and **sterile jacket** of the antheridium. Know that **sperm nuclei are produced by mitosis.** Recognize the following parts of the archegonium: **egg, venter, neck.** Know that **egg nuclei are produced by mitosis.** In the sporophyte recognize the **calyptra, sporangium, stalk. Know that spore nuclei are produced by meioses.**

IIIa. View the Colonies on Demonstration

Most mosses are very similar in form. Most have gametophytes consisting of a central axis with leaf-like structures attached. The structures that appear to be sporangia are actually the sporophytic generation.

IIIb. Hairy Cap Moss

Take a petri dish from the front with preserved hairy cap moss. Of all our native mosses this genus is the largest. It is ideal for your study of basic moss morphology. Note that this species has separate male and female gametophytes. Each petri dish contains, as a minimum, one male plant with a splash platform where the antheridia are located, and two female plants each with a sporophyte, one of which has a calyptra and one without. Please leave one sporophyte with the calyptra intact for others to view!

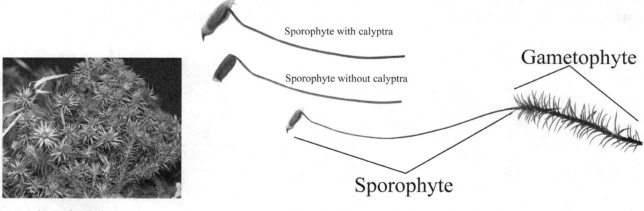

Male *Polytrichum* Moss   Female Gametophytes with Sporophytes

IIIc. Prepared Slides of Mnium Gametangia

Mnium moss is similar to hairy cap moss in that it has separate male and female gametophytes. The antheridia on the male plants are also clustered into splash platforms as observed in hairy cap moss. Hence use the material from "IIIb" as a reference while viewing these slides.

Mnium: Antheridium

Take the slide labeled "*Mnium*: Antheridium" and hold it to the light and compare with the male plant of hairy cap moss. Note the slide is a longitudinal section through the central axis of the tip of the male gametophyte. Place the slide under your microscope and observe the antheridia.

Draw an antheridium; label spermatogenous tissue and sterile jacket layer.

Mnium Archegonium

Take the slide labeled "*Mnium* Archegonium" and hold it to the light. Note the slide is a longitudinal section through the central axis of the tip of the female gametophyte. Place the slide under your microscope and observe the archegonia. Note that because of the length of the archegonium it is next to impossible to see one intact. You will either see a venter with an egg without a neck, or, towards the top, necks without venters.

Make a composite drawing of an archegonium; label venter, egg and neck.

IIId. Moss Protonemata

When moss spores germinate they form a filamentous structure called a **protonema** (plural **protonemata**). Protonemata are like branched filamentous green algae except that the cells have numerous disc shaped chloroplasts.

Make a wet mount of the liquid culture of germinated hairy cap moss spores and observe with your microscope.

Draw a moss protonema.

Discussion Topics

A **clade** is a group of organisms that share a set of common characteristics due to direct linear decent from a common ancestor that had those characters. A **grade** is a group of unrelated organisms grouped due to their superficial similarities, often because the members are rather simple in structure.

1. Is the Kingdom Protista a grade or a clade? Why ?

2. Are the bryophytes (not Bryophyta) a grade or a clade? Why?

3. What selective pressures work to keep plant gametophytes small?

4. In all three phyla of bryophytes, we saw adaptations to elevate the mature sporophyte. Describe all three. Why is it adaptive for the sporophytes to have greater stature?

5. Consider the cladistic diagrams on page 7. Neither are complete or correct. Create a better diagram. With the information at hand can we be certain about the sequence of events involving the evolution of the three phyla of bryophytes and how it relates to the vascular plants?

6. From the readings in your text, did the green algal progenitor of the plants have an alternation of generations? How could the factors considered in questions 3 & 4 have affected the life cycles of the earliest plants?

Studying Plant Kingdom: Seedless Vascular Plants of Phyla Pterophyta and Lycophyta

INTRODUCTION TO VASCULAR PLANTS

Other than the three phyla of bryophytes all plants have vascular tissues. As discussed earlier, the mosses (phylum Bryophyta) have cells that serve to conduct water and photosynthate, and these cells are homologous to vascular tissues. True vascular tissues, however, include xylem with tracheary elements. Like the hydroids in the mosses, these cells are dead at maturity and serve as pipes through which water moves by mass flow. Unlike the hydroids, however, the tracheary elements have secondary walls impregnated with lignin. This substance make tracheary elements (and other cells with secondary walls) strong, rigid and self supporting. The presence of lignin allows vascular plants to grow tall and erect.

In all vascular plants the sporophytic generation is the dominant form, and only sporophytic plants have vascular tissues. As we go from the non vascular plants to the primitive vascular plants to the seed plants, we observe an evolutionary tendency where the gametophye becomes progressively reduced relative to the sporophyte. In the angiosperms, this reduction is so complete that without a microscope we would not recognize these plants as having an alternation of generations at all. In the following exercises we want to consider this trend.

THE FERN ALLIES

The fern allies are all the vascular, non-seed plants other than the ferns. They are not a natural group and are grouped into three different phyla.

> **Domain** Eukarya
>
> **Kingdom** Plantae
>
>   The Fern Allies
>
> **Division** Psilotophyta
>
> **Division** Lycophyta
>
> **Division** Sphenophyta

I. DIVISION: PSILOTOPHYTA

One Example *Psilotum*

Objectives:

As for all plants, know that there exists an alternation of generations consisting of sporophytic and gametophytic generations; that gametes are produced by mitosis and spores by meiosis. As for all vascular plants, the sporophyte alone has vascular tissues and is the more obvious member of this duality. Recognize *Psilotum* sporophytes to genus, phylum, and king-dom Recognize that the sporophyte of *Psilotum* has no true roots or leaves and has only one plant organ, the stem. Know that pattern of branching of the sporophyte.

On the side bench we have living material of sporophytes of *Psilotum nudum* commonly called the whisk fern. Observe the plant and determine the pattern of branching.

How does branching occur in the sporophyte of *Psilotum*?

You have seen this pattern with two other organisms previously studied. Which two are they?

Speculate: is either of these occurrences an example of homology with that of *Psilotum*?

Note: the shoot consists of only one plant organ, the stem. While there are small structures coming off the stem (enations), these are not vascularized and, hence, are not leaves. Now observe the de-potted plant material. Note that the belowground portion of the plant is also made up of only stem material. While it is unlikely that *Psilotum* is truly a direct descendent of the Rhyniophytes preserving their primitive structure, the structure of *Psilotum* is much like that of these earlier plants.

If you have not already identified the sporangia on *Psilotum,* do so now. Observe the demonstration slide of the sporangium.

What nuclear division process resulted in the production of the spore nuclei?

II. DIVISION: LYCOPHYTA

Objectives:

Vegetative Structures: Students should know that if the Psilophytes are considered to be secondarily reduced, members of this phylum are the most primitive vascular plants. As in all vascular plants the sporophytic generation is the most prominent. Sporophytes of this group have true leaves, roots and stems. As in the Psilophytes, branching occurs either as equal or unequal dichotomies. Leaves of this group are not considered homologous with the leaves of other plants. They are microphylls with only one vein, and not associated with a leaf gap (we will explore this when we get to the ferns).

Sexual Structures: Spores are borne in **sporangia** borne on leaves termed **sporophylls.** In most cases sporophylls are clustered at the tip of the shoots forming cones termed **strobili.**

In the case of the club mosses (family Lycopodiaceae) only one type of spore is produced resulting in the production of bisexual gametophytes. This condition is termed **homospory** and the club mosses are **homosporous.**

In the case of the spike mosses (genus *Selaginella*) Two types of spores are formed. One type is larger and results in the formation of only female gametophytes. These larger spores are termed **megaspores**, and the female gametophytes resulting from their germination are termed **megagametophytes**. Megaspores are borne in **megasporangia**, which are borne on **megasporophylls**. Another type of spore is smaller and results in the formation of only male gametophytes. These smaller spores are termed **microspores**, and the male gametophytes resulting from their germination are termed **microgametophytes**. Microspores are borne in **microsporangia**, which are borne on **microsporophylls**. In the case of *Selaginella*, megasporophylls and microsporophylls are borne in the same strobilus.

Names: Know the common name of members of the Lycopodiaceae, **club mosses**, and that they are homosporous. Know the genus name of the spike mosses, ***Selaginella***, and that it is heterosporous.

IIa. The Club Mosses

Vegetative Morphology: Observe the living examples of the club mosses on the side bench. Carefully verify that the pattern of branching in all three cases involves the dichotomous splitting of the apical meristem. Note the sporangia and the sporophylls on the non strobilus forming, *Huperzia lucidulum*. The sporophylls of this plant alternate with leaves that are associated with asexual reproductive structures called bulbils. Note if you see strobili on either of the two other plants. Also observe the pressed specimens, especially note the rhizomes with roots attached.

Prepared Slide of Strobilus of Lycopodium: Take this slide and hold it to the light. It was made from a longitudinal slice through a strobilus (compare with strobili on the living and/or pressed plants). Place the slide under the microscope and look for the following: the central stem serving as the axis for the structure; the **sporophylls** bearing the **sporangia**, the sporangia and the **spores** all of which are the same size. Label the diagram below:

A = _____

B = _____

C = _____

Gametophyte: See the demonstration on the side bench of the club moss gametophyte (*Lycopodium*) with attached young sporophyte. This particular gametophyte is non photosynthetic being dependent on fungi in the soil for nutrition.

Since *Lycopodium* is homosporous which sexual structures were borne on this gametophyte?

IIb. Selaginella

Observe the living examples of Selaginella on the side bench. Take a strobilus from the side bench and take it to your seat. Now take the prepared slide of the longitudinal section of a *Selaginella* strobilus and compare that section with the whole strobilus. Place the slide under the microscope and look for the following: the central stem serving as the axis of the structure; **microsporophylls** bearing the **microsporangia** with **micro-spores**; **megasporophylls** bearing **megasporangia** with **megaspores**. Label the diagram below:

A = _____

B = _____

C = _____

D = _____

E = _____

F = _____

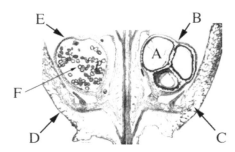

Optional: If you have time and interest dissect the preserved strobilus you took earlier and identify the structures in the diagram above. Otherwise return this strobilus to the stock bowl.

Gametophyte: Observe the demonstration of the megagametophyte with attached young sporophyte on the side bench. Note that it develops entirely within the old megaspore wall.

A = Megagametophyte inside old megaspore wall.

B = Microspore

C = Root of young sporophyte

D = Shoot of young sporophyte

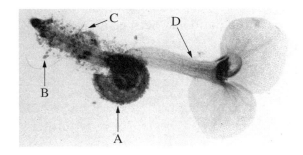

III. DIVISION SPHENOPHYTA - THE HORSETAILS

Objectives:

Students should recognize *Equisetum* to kingdom, phylum and genus.

While the horsetails are structurally more complex than the other two phyla of fern allies, they are an ancient group. *Equisetum* is structurally identical to fossil materials over a hundred million years old. Observe the living and pressed specimens of *Equisetum* on the side bench. These sporophytes have well defined nodes and internodes. At the internodes whorls of leaves are attached. These leaves are small and non-photosynthetic. Like higher plants, the nodes are the site of lateral bud formation. In this case, however, the buds do not reside in the axils of the leaves, but below the leaves. Some species form whorls of lateral shoots from these buds so observe the diversity of form available. Of the three phyla of fern allies, the Sphenophyta in the form of the genus *Equisetum*, is the most common. Horsetails can be found in sour pastures and is considered a weed in that it destroys B vitamins if eaten by livestock. Another species of Equisetum is frequently found along railroad tracks in Wisconsin. Note the demonstration of the strobili of Equisetum on the side bench. These strobili are considerably different from those of the Lycopods. It consists of a central stem to which modified stems (sporangiophores) bearing sporangia are attached.

Discussion Topics

Are the fern allies a clade or a grade?

Heterospory evolved independently at least four times. What is an adaptive advantage of heterospory?

From an evolutionary perspective, why don't we find vascular tissues in the gametophytes of the vascular plants?

Club mosses are unimportant ecologically and economically. Why should they be protected?

Based on the structure of the gametophyte and on molecular studies, some botanists hypothesize that *Psilotum* is a type of fern. This would mean that its morphology is secondarily derived. Its structure is simple but is it primitive? What is the difference between simple and primitive?

IV. PTEROPHYTA - THE FERNS

Ferns are the most widespread and ecologically important of all the seedless vascular plants. The fern sporophyte is structurally complex with true stems, roots and leaves. The leaves are macrophylls in that they have numerous veins and are associated with a leaf gap (an area of parenchyma in the stem where the vascular trace diverges from the stem's vasculature). While some ferns have non photosynthetic gametophytes and some are heterosporous, the gametophytes we will see in lab are photosynthetic and are of homosporous species (hence, these gametophytes are bisexual).

Objectives:

Recognize examples to phylum and recognize **sporophytes** and **gametophytes.** For the sporophyte recognize the **leaf** (frond), and the different parts of the leaf (**petiole, blade, and sori**), the **stem** (in many cases a rhizome), and the **roots**. From the gametophyte recognize **archegonia**, and the parts of the archegonium (**neck, venter, egg**) and the **antheridium** and its parts (**spermatogenous tissue, sterile jacket layer**).

Domain Eukarya

Kingdom Plantae

Division Pterophyta

V. THE SPOROPHYTE

Take a pressed specimen from the front bench to your seat.

Identify and include in a labelled drawing the following parts: frond, petiole, blade, sori, rhizome (stem), roots.

Observe the living sporophytes in the room and locate the stem and the leaves.

Va. The Stem

Take the prepared slide of the Rhizome of *Dicksonia* and hold it up to the light. Note, this is a cross section through a stem of this fern. Place the slide on your microscope and observe the cylinder of red stained cells. This tissue is xylem and the stain binds with lignin. Note the dense tissue in the very center of the stem is not xylem. On either side of the xylem is a layer of phloem. Label the figure on the next page.

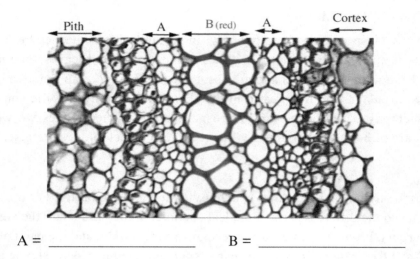

A = _____   B = _____

Vb. The Leaf

Look at the demonstration slide of a leaf gap. This is definitive of macrophylls which is the type of leaf ferns have. Take the prepared slide labelled sorus of *Cyrtomium* and place it on your microscope. This is a cross section through the blade of *Cyrtomium* (holly Fern). Note the numerous veins throughout the leaf, this is also characteristic of macrophylls. Now carefully study a sorus with sporangia.

Are the spores all one size?

What nuclear process resulted in the production of the nuclei in these spores?

Based on your observation here, will these spores produce a unisexual or bisexual gametophyte?

VI. GAMETOPHYTE

Take a living fern gametophyte available at the front and place it on a microscope slide in a drop of water. Do not add a cover slip yet. Study this plant either with your compound microscope or with a dissecting microscope. The plant is thin and translucent. Note the complete absence of vascular tissue. Add a cover slip and observe at 100x and 400x. Note the rhizoids that anchor the plant to its substrate. Also note the distinctive disk shaped chloroplasts of the photosynthetic cells.

Does the growth form of the gametophyte remind you of any other group of plants studied previously? Which one/ones?

Now take the prepared slide of a whole-mounted gametophyte and look for gametangia. Archegonia can be found near the notch and antheridia will be found scattered across the whole thallus. The archegonium is embedded in the tissue of the gametophyte. To observe the egg in the venter you must carefully through focus.

Make two drawings of an archegonium (400x). One with the focus on the neck, the second with the focus on the venter. Label the neck and the venter.

What type of nuclear division resulted in the formation of the egg nucleus?

Make a drawing of an antheridium (400x).

Draw the entire gametophyte at 40x. Label archegonia and antheridia.

VII. GAMETOPHYTES WITH ATTACHED YOUNG SPOROPHYTES

Take a petri dish containing a gametophyte with attached young sporophyte.

Draw these two generations: Label gametophyte and sporophyte.

Discussion Questions

1. On volcanic islands, ferns are often the first vascular plants to appear. How can you explain this.

2. Bracken ferns are important on many dry sites here in Wisconsin. How can they persist if they are dependent on liquid water for fertilization?

3. Researchers have successfully produced diploid gametophytes. Surprisingly they do not possess vascular tissue. They produce gametes that result in tetraploid sporophytes. If the number of sets of chromosomes is not the deciding factor in the morphogenesis of these pants, can you think of other factors that influence the proper development of the sporophyte?

4. In the Appalachians there are species of fern that do not have a sporophytic stage. The gametophytes reproduce vegetatively only. In terms of survival, how could this possibly be beneficial to these species?

Studying Plant Kingdom: Gymnosperms of Phyla Cycadophyta, Ginkgophyta, Coniferophyta, and Gnetophyta

Domain Eukarya

Kingdom Plantae

    Division Coniferophyta - The Conifers

    Division Cycadophyta - The Cycads

    Division Ginkgophyta - The Ginkgoes

    Division Gnetophyta - The Vessel Bearing Gymnosperms

SEED PLANTS: AN OVERVIEW OF TERMS

The remaining five phyla of plants to be studied this semester are all seed plants. Seeds are borne on sporophytes and contain young sporophytes (embryos). Superficially it appears as if these plants skip the gametophytic generation. In all cases, the gametophytic generation is still there, but it is either hidden by sporophytic tissues or reduced. It is only our knowledge of the life cycles of non seed plants that give us a perspective in which to understand the evolution and life cycles of the seed plants.

All seed plants are **heterosporous**. All the terms we introduced in regards to the *Selaginella* life cycle is still pertinent. Below is a list of old terms and new ones necessary to understand seed plant life cycles. You need to know and understand then all.

Micro: Suffix that originally denoted small but has, in a botanical sense, taken on the meaning "male".

    Microspore Mother Cell = Microsporocyte: Diploid cell destined to undergo meiosis to produce microspores in the seed plants.

    Microsporangium: Sporangium that bears microspores. In seed plants synonymous to a pollen sac.

    Microspore: Spore that develops into a microgametophyte.

    Microgametophyte: In seed plants this is the pollen grain.

Microsporophyll: Modified leaf that bears microsporangia. Examples from the seed plants include the stamens of flowers and the subunits of the pollen cones of conifers.

Microsporangiate cone = male cone: terminal clusters of microsporophylls such as the pollen cones of conifers and cycads.

Mega: Suffix that originally denoted large but has, in a botanical sense, taken on the meaning "female"

Megaspore Mother Cell = Megasporocyte: Diploid cell destined to undergo meiosis to produce megaspores in seed plants. This cell is buried in the megasporangium.

Megasporangium: In seed plants is called the **nucellus.** Is surrounded by sterile tissues called **integuments**. The integuments retain both the nucellus and the megagametophyte keeping this generation bound to the sporophyte.

Integuments: Sterile tissues surrounding the nucellus. These tissues mature into a seed coat.

Megaspore: haploid cells resulting from the meiosis of a megasporocyte. It develops into the megagametophyte. In seed plants the megaspore is not released but is retained in the nucellus.

Ovule: The integuments together with the nucellus form the ovule. Later stages include the **megagametophyte** and embryo.

Megagametophyte = Female gametophyte. In the seed plants is retained on the sporophyte in the nucellus. In all the gymnosperms is present in the mature ovule (seed) where it functions as the food storage tissue for the embryo. In the angiosperms this generation is destroyed by **double fertilization**.

Megasporophyll: Modified leaf bearing the megasporangia (hence the ovules). Not all seed plants have them. Cycads are one group with megasporophylls. In flowering plants megasporophylls are termed **carpels**.

Megasporangiate cone: Structure composed of subunits bearing **ovules**. These subunits are not megasporophylls but seed-scales in the conifers.

Seed: Mature ovule bearing a young sporophyte (an embryo) complete with food storage tissue covered by a seed coat derived from integuments.

THE GYMNOSPERMS

The gymnosperms, like the bryophytes, are no longer recognized as a taxonomic category. There are four phyla of seed plants historically considered in the group. One phylum, **Gnetophyta**, has clear links to the **Anthophyta** (flowering plants), the one seed plant phylum not considered a gymnosperm. The relationship of the other three to each other and to the flowering plants is unclear. It is even possible that the seed habit evolved more than once, and that the common ancestor of these four phyla was not a seed plant.

All gymnosperms are seed plants, but their only other point of similarity is that they are not flowering plants. Flowering plants are partly defined by their megasporophylls which enclose the ovules (**carpels**). Gymnosperms do not enclose their ovules and are commonly known as being **naked-seeded**. The other point of commonality setting these groups setting them apart from the flowering plants is that the nutritional tissue included in the seed is the **megagametophyte**.

Learning Objectives: You should recognize all examples seen during class to phylum. In the case of the Coniferophyta, you need to also recognize the genus, *Pinus,* and in the case of Ginkgophyta the species *Ginkgo biloba.* For the Coniferophyta you need to recognize and be able to name all the structures listed in the introduction, *Seed Plants: An overview of Terms* as demonstrated by the materials studied in class. Further you need to also recognize other structures unique to the Coniferophyta (seed scale, sterile bract, seed scale complex, the ovulate cone consisting of a shoot with spirally arranged seed-scale complexes.

I. DIVISION CONIFEROPHYTA - THE CONIFERS

Phylum of seed plants all of which become woody. The life cycle does not include motile sperm. The sperm nucleus is carried to the egg solely by the pollen tube. While lacking in the yews, the most distinctive feature of the group is the ovulate cone, consisting of a central stem bearing **seed-scale complexes**. Each seed scale complex consists of a modified leaf called a **sterile bract**. In the axil of the sterile bract a modified stem called a **seed scale** develops. This in turn bears the ovules.

Ia. Demonstration of Conifer Diversity

Conifers include many commercially and ecologically important species. These include the pines, spruces, hemlocks, douglas fir, junipers, cedars and all three genera of the redwoods. The world's largest, tallest and oldest trees are all conifers. Carefully observe the examples of conifers on display. All the living examples were collected from campus. Note the diversity of form. You should be able to recognize all these examples to phylum. Pay special attention to *Pseudotsuga*. Its mature cones clearly show the sterile bracts associated with the seed scales of the ovulate cone. **Do not damage the large pine cones (Coulter and Sugar pines) or the redwood cones!** These are western species and these cones cannot be easily replaced.

Ib. Genus Pinus

Pinus resinosa (red pine) is a commonly planted species of pine in Wisconsin. We study it in detail as our primary example of the conifers.

Identifying the Genus Pinus: The genus *Pinus* is unique in that all the leaves of mature plants are born in fascicles. The fascicle of pine is a dwarf shoot that bears the leaves and dies when the leaves die. Observe the red pine boughs in the lab room.

How many leaves are borne on a fascicle in *Pinus resinosa*?

Microsporangiate Stages: In conifers microsporangiate cones are borne for only a matter of days or weeks and are then shed. In *Pinus* the microsporangiate cones are borne in clusters and emerge with the new spring growth of the lower branches. Here in Dane county the microsporangiate cones emerge in May.

Clustered Microsporangiate Cones

Observe the clusters of microsporangiate cones on the boughs at the front of your bench.

Explain the alternating bare areas interspaced by leafy areas behind the clustered cones:

Studying the Details of the Microsporangiate Cone

Take a microsporangiate cone from the bowl next to the bough bearing the clustered cones. Take the prepared slide labelled, "*Pinus* Male Strobilus". Study the whole cone and then compare with the section using your unaided eye. You should be able to correlate that the slide is indeed a longitudinal section through a male cone.

Study this prepared slide. Identify the **microsporophylls, microsporangia** (pollen sacs) and the **immature microgametophytes** (pollen grains). **Label the figure.**

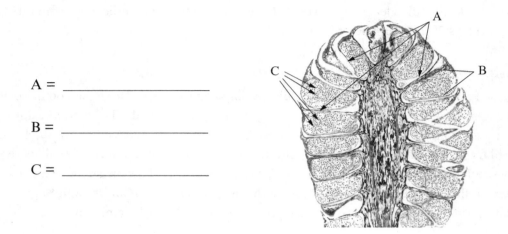

A = _____

B = _____

C = _____

After viewing the prepared slide dissect a whole cone. Use a dissecting microscope! This is best done by initially pulling the cone apart using your fingernails then using teasing needles to dissect out individual mircrosporophylls.

Make two drawings of the same microsporophyll. Label one, "upper surface", and the other, "lower surface." Also label the pollen sacs and the microsporophyll itself.

The Microgametophyte (Pollen Grain)

In all seed plants the microgametophyte is greatly reduced. For fertilization to occur the microgametophyte must be physically carried to the ovule. This event is **pollination** and in pine is accomplished by the wind. Sample the water at the bottom of the dish with the microsporangiate cones with a pipette and make a wet mount.

Draw a pollen grain.

What possible survival advantage could be provided by the "Mickey Mouse" ears?

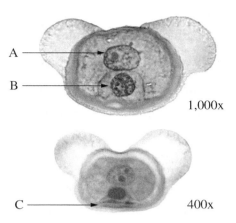

1,000x

400x

Again view the prepared slide under your microscope and look carefully at the pollen grains in the pollen sacs at 400×. Identify the nuclei of the tube (A) and generative (B) cells. You should just be able to see the boundary between these two cells. Also, by looking at a number of different pollen grains you should be able to discern the prothallial cells (C) though they may not appear as two distinct cells. These are all that is left of the vegetative body of the microgametophyte. The tube cell will germinate into the pollen tube which will grow through the nucellus and eventually up through the neck of the archegonium to an egg. After pollination the generative cell will divide to form a stalk cell and a spermatogenous cell. The spermatogenous cell, in turn will undergo mitosis to form two sperm nuclei. Both of these nuclei will be delivered to the egg nucleus via the pollen tube.

Megasporangiate Stages

In conifers megasporangiate cones persist through the year and their presence give rise to the common and phylum names of the group (the conifers - Coniferophyta). In *Pinus* the megasporangiate cone emerges with the new spring growth of the upper branches. Here in Dane county the megasporangiate cones emerge in May and do not reach maturity where they disperse seed until the fall of the following year.

Bough with Megasporangiate (ovulate) Cones

Study the pine bough at the front bench with the various stages of ovulate cones. The smaller cones emerged this year in May and are six months old. The older cones emerged the preceding May and are 18 months old. Note, there may be even older cones still attached to the bough. If so these are three-years old and have been dead for a year. In pine the ovulate cones are not actively abscised, but eventually fall off due to the growth of the bark and the wood.

Studying the Details of the Megasporangiate Cone

Take a whole six month old cone to your seat and compare it with the prepared slide labeled "Pine: female strobilus". Verify that the section is indeed a longitudinal section. Observe the prepared slide through your microscope. Identify the sterile bracts labelled "A", and the seed scales labelled "B" on the next page. A sterile bract with its associated seed scale is a seed scale complex. All conifers except the yews have ovulate cones made up of seed scale complexes though they are reduced in *Podocarpus*.

Dissection of a Megasporangiate Cone: Work with a partner. Return one six month cone to the stock bowl. Take the other cone and pull it apart using your fingers. Give part to your partner. Each of you should then dissect out one seed-scale complex. Use a dissecting microscope! Lay it flat and observe the ovules.

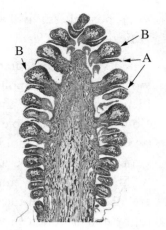

How many ovules are there on each seed-scale? _____

Flip over the seed-scale and observe the sterile bract.

Draw each view of one seed-scale complex. Label **seed-scale, ovules**, and **sterile bract**.

Observe a mature open cone. Note that the sterile bracts have not grown in step with the seed-scales. Observe the douglas fir cone at the conifer display to see an example where the sterile bracts are visible in a mature cone.

Studying the Details of the Ovule

Take the prepared slide labeled "Pine female strobilus"and use 400x to study details of the ovule. To observe all the structures listed you will need to view a number of different ovules as it is rare to find one section with all these structures. Identify the **integument**, and the **micropyle** of the ovule where the integument comes together but fails to meet. Look for an ovule with a huge cell in the center. This cell is the **megaspore mother cell**. It will undergo meiosis to produce four **megaspores**. In *Pinus,* only one megaspore will go on to produce a **megagametophyte**. The tissue in which this cell is embedded is the **nucellus (megasporangium)**. It is surrounded by the integument, but note that the two tissues are not joined near the micropyle. The space between these two tissues is the **pollen chamber**. Carefully search the pollen chambers of several ovules for pollen.

Has pollination occurred?

How does pollen get to the ovule and then into the pollen chamber?

Label the figure of the ovule below. Note the micropyle is not in the plane of section.

A = _____

B = _____

C = _____

D = _____

E = _____

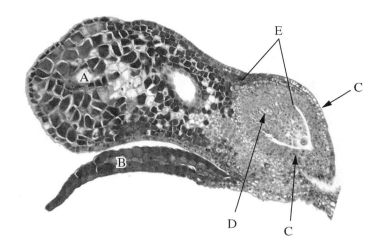

Studying the Megagametophyte

Take the prepared slide labelled "Pinus Archegone" and observe the slide with your unaided eye. Note that this is a longitudinal section through an ovule. The megagametophyte occupies the central region of the ovule and one or two eggs will be visible without magnification. Observe the slide with your microscope. Note that the eggs are associated with an **archegonium**. It is improbable that you will have a section that perfectly sections the neck, however the cells of the venter should be easy to identify at 400×. Identify the following: **megagametophyte, venter** of the archegonium, **egg, nucellus, integument. Label the Figures on this page.**

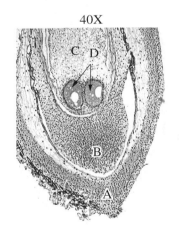

40X

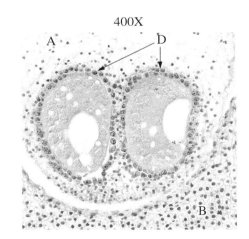

400X

A = _____   A = _____

B = _____   B = _____

C = _____   C = _____

D = _____   D = _____

Studying the Seed and Seedling Stages of Pine

Go to the front bench and take a Piñon pine seed. These seeds are called pine "nuts" and are used in pesto. They are a western species and these seeds are many times larger than the seeds of red pine. Make a small cut in the seed coat with a razor blade, then using your finger nails peel off the seed coat.

What tissue gave rise to the seed coat? _____

Immediately below the seed coat is a papery brown layer. Peel this layer off also.

Based on your understanding of the pine ovule, from which tissue was this brown layer derived?

_____

The white tissue exposed is female gametophyte. To expose the embryo, dig your thumb nails into the tissue and rip the structure in two. This should clearly reveal the embryo inside. Identify the cotyledons and the central axis of the stem and root.

Take the slide "Pinus: embryo", hold it to the light and study the section with your unaided eye. Note this is a longitudinal section through a pine seed that has had the seed coat removed. Observe this slide with your microscope. Identify the **cotyledon**s (seed leaves), the **apical meristem of the shoot**, and the **apical meristem of the root**. Label the figure below:

Figure of Dissected Pine Seed (right) and longitudinal section (left).

A = _____

B = _____

C = _____

D = _____

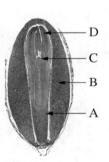

Seedling Stages: Study the demonstration of the various seedling stages at the front.

What is the function of the cotyledons immediately after germination?

II. DIVISION CYCADOPHYTA

Seed plants with non-woody secondary growth. Plants are dioecious either bearing megasporophylls or microsporophylls. Foliage is palm-like with pinnately compound leaves. Cycads have flagellated sperm carried to the archegonium by means of the pollen tube. This group was more important ecologically in the past than in the present. Today the group consists of only three families of plants. Cycads are extensively planted as ornamental.

Learning Objectives: Recognize examples by the common name **cycads** and to the phylum Cycadophyta.

Observe the examples of Cycads on the side bench. Note the demonstration of Microsporangiate and megasporangiate stages for both *Cycas* and *Zamia*.

Zamia has both **megasporangiate strobili** consisting of **megasporophylls** bearing **ovules**, and **microsporangiate strobili** consisting of **microsporophylls**. All sporophylls are organized into determinate strobili (strobili consisting of a shoot where the apical meristem does not persist, hence the strobilus terminates growth at that point).

Cycas has its microsporophylls arranged as above. However, the megasporophylls are not organized in determinate strobili. They form at the apex of the shoot and are followed by new growth consisting of vegetative leaves.

III. DIVISION GINKGOPHYTA

Seed plants with woody secondary growth and distinctive fan shaped leaves. Leaf of *Ginkgo biloba* Plants are dioecious. The males bearing microsporangiate strobili, the females ovules on peculiar stalk-like structure each with two ovules. The ovules mature into a stinking fruit-like seed. *Ginkgo* has flagellated sperm carried to the archegonium by means of the pollen tube. The group was more important ecologically in the past than in the present. Today the phylum includes one species, *Ginkgo biloba*, which is extensively planted as a street tree.

Leaf of ***Ginkgo biloba***

Objectives:

Recognize *Ginkgo biloba* to species and phylum. Recognize its fruit-like seed as a seed and not a fruit.

Observe the demonstration materials of *Ginkgo biloba* on the side bench.

From what tissue is the fleshy covering of the *Ginkgo* seeds derived?

IV. DIVISION GNETOPHYTA

This phylum is commonly called the vessel-bearing gymnosperms because, like the flowering plants, and unlike the other three phyla of gymnosperms, their xylem contains vessel cells. This is a peculiar group consisting of three genera that are radically different in shape and form from each other. These plants are dioecious. The Gnetophyta is thought to be the group of gymnosperms most closely related to flowering plants.

Objectives:

Recognize the three genera to phylum.

Your TA will take you to the greenhouse to view these plants.

Label each figure below with the appropriate genus name (*Ephedra, Gnetum, Welwitschia*):

Genus: _____

Genus: _____

Genus: _____

Studying Plant Anatomy: The Angiosperms

The angiosperms (flowering plants) were the last phylum of plants to appear on the Earth. The earliest unambiguous flowering plant fossils are from the Cretaceous the last period of the Mesozoic. The group diversified explosively and many genera found today are represented early in their fossil record. There is a great deal of uncertainty about the primitive characters of the group. All sexually reproducing flowering plants have flowers. The general rule is that the flower is a determinate shoot consisting of four types of modified leaves, **sepals, petals, stamens** (microsporophylls), and **carpels** (megasporophylls). However there are major groups of flowering plants that have flowers without petals or sepals. Examples include the oaks and birches. It is not clear whether these trees descended from plants that never had petals and sepals, or if their floral structure has become reduced as an adaptation to wind pollination. The grasses are another group adapted to wind pollination. Their flowers typically have only stamens and carpels. In the case of the grasses, however, it is clear that this represents a reduction in complexity from ancestors that did have petals and sepals. Today the field of systematics is in ferment. Taxonomists are searching for ever older examples of flower fossils, and are evaluating the molecular evidence provided by living groups in an attempt too gain a clearer picture about both the evolution of the group from some gymnosperm ancestor, and the patterns of evolution within the group. One thing is clear. The angiosperms are the most abundant and diverse group of plants on Earth.

Defining characteristics: All members of the phylum Anthophyta have flowers. At a minimum, flowers include carpels and/or stamens. The carpel is a megasporangium that bears and encloses ovules. After pollination, carpels develop along with the enclosed ovules to form a **fruit**. Further, angiosperms all have **double fertilization**. The microgametophyte produces two sperm nuclei which are delivered to the megagametophyte via the pollen tube One fertilizes the egg to produce a zygote. The other undergoes fusion with the two nuclei of the central cell givinf rise to the **endosperm** tissue.

I. FLOWERS

Ia. Magnolia Flower

Take a bowl with a dissected Magnolia flower from the front desk to your seat. To aid you in this study, your TA has a picture of the flower in its living form displayed on the teaching monitors. The central axis of the flower is termed the **receptacle**. The receptacle bears four types of modified leaves. At the base of the flower occur two whorls of petal-like structures. The lowest whorl is termed the **calyx**.

The individual members of the calyx are the **sepals**. The next whorl up is termed the **corolla** and its individual members are the **petals**. Above the petals are born the stamens which together make up the **andreocium**. Note that these are not whorled but spiralled around the receptacle. To see this pattern study the scars on the receptacle where the stamens were detached. This arrangement is considered a primitive characteristic in flowers.

Take a detached stamen and observe it using a dissecting microscope. Note the pollen bearing sacs termed the **anthers**. These are microsporangia and the stamen is a microsporophyll. Also note the stalk below the anthers that attach the stamen to the receptacle. This is the filament of the stamen.

Draw the stamen of *Magnolia* - label anthers and filament.

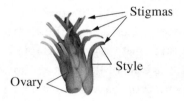

The Gyneocium: The structures above the andreocium are carpels which together make up the gyneocium. Like the stamens these are also spiralled around the receptacle. The carpels in *Magnolia* each represent a single **pistil**. Pistils are made up of one or more carpels. Each pistil includes three parts. The base where the ovules are contained, is termed the **ovary**. The tip where pollination occurs is termed the **stigma**, and the part between the two is termed the **style**. Because the pistils are positioned above Gyneocium of the other three parts of the flower (calyx, corolla, and Magnolia andreocium) they are termed **superior**.

Label the figure:

A = _____

B = _____

C = _____

D = _____

E = _____

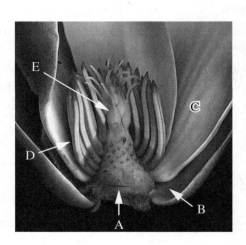

Other notes: Observe the non-dissected flower at the front situated face up. Note that these flowers are **radially symmetrical.**

Magnolia does not produce nectar. What other reward might lure pollinators to the flower?

Based on your observations of the structure of the flower, does *Magnolia* limit access to any potential pollinator?

Ib. The Snapdragon Flower

We have plenty of material and you should dissect two flowers. These are at the front bench.

First Dissection

Observe an intact flower. Note that the flower is **bilaterally symmetrical.** Which floral parts are fused?

Grasp the corolla and gently pull it off the flower. The pistil should be clearly visible. Is the pistil superior or inferior?

Now take the corolla tube and cut a slit through its side. Lay the tissue flat in a single layer. Note that the stamens are fused to the corolla.

The Pistil: Observe the pistil:

Make a drawing - label ovary, style and stigma.

Dissect the ovary by cutting it in two with a cross sectional slice. Observe the internal structure of the ovary. Identify the **ovules**. The openings in the ovary are called **locules**. Often there is a locule for every carpel in a compound pistil (remember the ovary is part of the pistil). Observe where the ovules are attached to the wall of the ovary. This tissue is termed the **placenta**.

Draw a cross section of the ovary. Label each carpel, the placenta, and the ovules.

Second Dissection

This time slice the flower in two using a sharp razor blade exactly along the plane of symmetry. Label the diagram on the next page.

A = _____

B = _____

C = _____

D = _____

E = _____

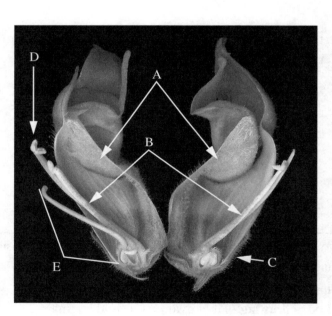

Overview: This flower has characters considered to be less primitive than those of Magnolia. These are outlined below:

Fused floral parts
    The petals are fused
    The stamens are fused to the corolla
    The sepals are fused
    The carpels are fused to form a compound pistil.

●   **The parts of the flower (calyx, corolla, andreocium) are arranged in whorls** (unlike Magnolia none of the parts are spirally arranged).

The flower is bilaterally symmetrical

Evolution in the flowering plants is often driven by the need to get pollen from one flower to another of the same species.

How is the snapdragon flower adapted to this task?

Why might it be maladaptive to allow easy access to the nectar of the flower to all comers?

●

Ic. The Fuchsia Flower

We have one flower for each student. At the front bench observe the orientation of the flowers on the plant, then take a flower from the bowl to your seat.

Observe the external characteristics of the flower.

What is the symmetry of the flower?

Which parts seem to be fused?

●

Because the ovary is positioned below the other parts it is termed **inferior**.

Dissecting the Flower

Carefully slice the flower in two longitudinally. Start with the ovary and work your way up one side and then down the other. Lay the two halves of the flower with the cut surfaces facing up.

Which parts are fused?

Draw this dissected flower. Label the following parts: pistil, stigma, style, ovary, sepals, petals, stamen, anther, filament.

Based on your observations of the flower what pollinates this plant?

Id. Other flowers

Complete vs. Incomplete: Magnolia, snapdragon and Fuchsia flowers all have the four basic parts of a generic flower: calyx, corolla, andreocium, and gyneocium. Flowers with all three are termed **complete**. Flowers missing one or more of these parts are termed **incomplete**.

> **Sugar Maple**: Observe the demonstration of the pressed specimen of sugar maple on the side bench. Its flowers have a pistil and stamens but lacks petals and sepals. It is adapted for wind pollination.

Perfect vs. Imperfect: While sugar maple was incomplete it had both male and female parts. This condition is termed **perfect**. Some flowers have either stamens or pistils. These flowers are termed **imperfect**. All imperfect flowers are also incomplete.

> *Begonia*: Observe the flowering Begonia plant on the side bench. Note that on the same plant are found flowers either with stamens or with a prominently ribbed, inferior ovary. *Begonia* has imperfect flowers. Note the other examples of imperfect flowers by the *Begonia* plant.

Monoecious vs Dioecious: Plants with imperfect flowers that have both male and female flowers on the same plant like Begonia, are termed **Monoecious**. Plants that have only male flowers or female flowers are termed **Dioecious**.

> *Salix:* Observe the herbarium specimens of willow. Note that one has male flowers while the other has female flowers. *Salix* is a dioecious species.

II. THE LILY LIFE CYCLE

IIa. The lily flower: Observe the intact and partially dissected lily flowers on the side bench.

Are lily flowers bilaterally or radially symmetrical?

Are the flowers perfect or imperfect (one flower has had the stamens removed)?

Are the flowers complete or incomplete?

Is the ovary superior or inferior?

IIb. The Young Lily Bud: Take the prepared slide in your box labelled "Lilium: Pollen Tetrad" and hold it up to the light. Study it with your unaided eye. Note that this is a section of an unopened flower bud. At this stage the anthers are in the same plane of section as the ovary.

The Immature Anthers

Observe the anthers in this section. Note that they contain tetrads of microspores (note the slide label is incorrect). Also note the innermost tissue layer. This is the tapetum and it is important in the development of the pollen grains. It is consumed as the anther matures.

Draw an anther: label microspores and tapetum.

The Immature Ovary

Observe the ovary at 40x.

How many carpels make up this pistil?

Observe the placenta with attached ovules. Note the huge cells inside each ovule. These are the **megasporocytes** (megaspore mother cells). These cells reside in the **nucellus** (megasporangium). Note that there are **two integuments** and that they are incomplete. After pollination they will grow to encase the ovule. Label the figures below:

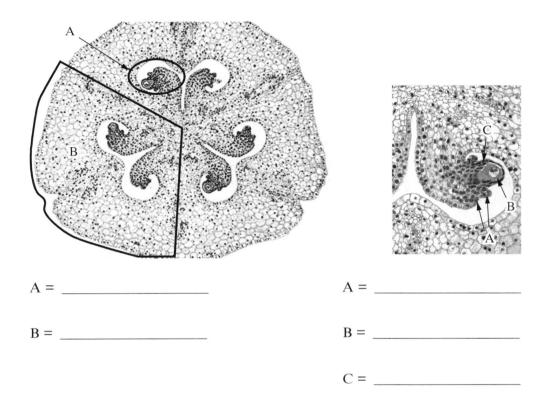

A = _____   A = _____

B = _____   B = _____

                            C = _____

IIc. The Immature Embryo Sac (megagametophyte): Take the prepared slide labelled "Lily Ovary" and observe it using your unaided eye. Verify that there are cross sections through a lily ovary on the slide. The three locules should be clearly visible in some of these sections. Observe the slide with your microscope. The four nucleated cell in the ovule is the **embryo sac** (the megagametophyte) at the four nucleated stage of development. Also identify the nucellus and the integuments.

Draw an ovule: label embryo sac, nucellus and integuments.

IId. The Mature Embryo Sac: Observe the pre-pared slide of the mature embryo sac on the side bench. This megagametophyte consists of only seven cells and eight nuclei. It will be destroyed by double fertilization but this will result in the formation of two genetically distinct cells, the **primary endosperm cell** and the **zygote**. The three cells opposite the micropyle are the antipodals (A). These seemingly have no function. In the center of the embryo sac is a huge binucleated cell, the central cell (B). This will undergo fertilization with one sperm nucleus resulting in **triple**

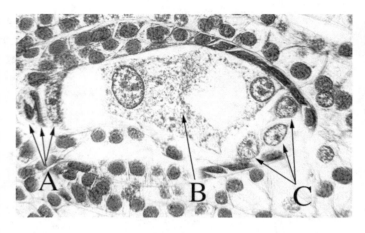

fusion of all three nuclei. The three cells closest to the micropyle are called the **egg apparatus**. One of these cells will undergo fertilization with one sperm nucleus resulting in a zygote. The other two will act as **synergids** and will be destroyed by the pollination tube.

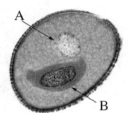

IIe. Dehiscing Anthers: Take the slide labelled "Lilium Mature Anthers" and study it with your unaided eye. Observe that this is a section made through an open lily flower where the anther is at the level of the style. Note the section of the style in the center surrounded by the anthers. Observe the slide with your microscope. Note the pollen grains (immature microgametophytes) inside each anther and the absence of the tapetal layer.

The Pollen Grain

Observe the pollen grains at 400x. The microgametophyte consists of a tube cell (A points to the nucleus of the tube cell) with a generative cell inside it (B). The tube cell germinates to produce the pollen tube. The generative cell undergoes mitosis to produce two sperm nuclei that are delivered to the mature embryo sac.

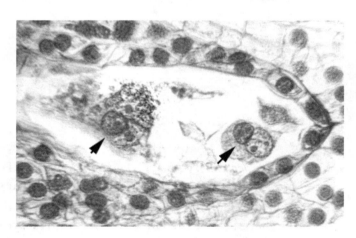

Double Fertilization: Observe the demonstration slide of double fertilization on the side bench. This is a difficult slide and may require through focusing. In the photomicrograph the right the two sperm nuclei are identified. The one to the right is undergoing karyogamy with the egg nucleus to form the zygote. The one to the left is undergoing karyogamy with the two polar nuclei to form the primary endosperm nucleus.

III. FRUITS

Pistils are made of carpels which are megasporophylls. The part of the pistil containing the ovules is the ovary. **Fruits are ripened ovaries**. One of the definitive characters of the angiosperms is that the ovules are encased. Fruits always serve to protect the maturing ovules/seeds. However, a great deal of variation exists in the angiosperms in regards to the nature of the mature fruit.

IIIa. Berries: Some fruits become totally fleshy. These fruits are berries. Many unrelated plants have fruits that are berries. In regards to **fruit-type** a great deal of convergent evolution has occurred.

● Observe the demonstration of berries at the front bench. For each, answer the following questions. How many carpels are represented?

How many locules are represented?

● **NOTES:**

What environmental factor drives the evolution of berries?

● ***IIIb. Drupes****:* Fruits that have a fleshy outer and a stony inner part are drupes. "Stone fruits" are drupes, and include cherries, peaches, plums, apricots and nectarines. All are members of the genus *Prunus* in the family Rosaceae. Other plants bear drupes as well but are unrelated. This is another example of convergent evolution.

Observe the demonstration of drupes at the front bench.

What environmental factor drives the evolution of drupes?

IIIc. Pomes: Pomes are a particular type of fruit borne by a group of plants in the Rosaceae including apple, pear, hawthorne and quince. In pomes the sepal, petal and stamen tissues have become fused to the ovary resulting in an inferior ovary. Pomes, like berries, are fleshy, but the tissues surrounding the ovary contributes the bulk of the flesh of the mature fruit.

NOTES:

IIId. Follicles: Follicles are dehiscent dry fruits derived from a simple carpel with one suture. The one obvious example familiar to many people are milkweed fruits.

NOTES:

How are these fruits adaptive?

IIIe. Legumes: Like pomes this fruit type is associated with a taxonomic group - the bean family (Fabaceae). This fruit is derived from one carpel with two sutures. Note that peanuts are not dehiscent but are still considered to be legumes.

NOTES:

Why have peanuts lost the ability to dehisce?

IIIf. Capsules: A capsule is any dehiscent fruit derived from a compound ovary.

NOTES:

IIIg. Grains, Nuts, Achenes: These are all examples of fruits that are non dehiscent and develop into a hard or stony tissue surrounding the seed.

Grains are the fruits of grasses (Poaceae). The carpel develops into a hard layer tightly fused to the seed coat. This covering is the bran stripped from the wheat kernel.

NOTES:

Nuts are fruits where the carpel becomes completely stony. Many fruits considered to be nuts, however, are actually drupes such as walnut, almond and coconut. Acorns are true nuts.

NOTES:

Achenes are hard fruits but not stony. An examples is sunflower "seed". Some achenes are winged and these are termed **samaras**.

NOTES:

IIIh: Aggregate Fruits: Some flowers have multiple pistils. Their fruits develop into aggregates of simple fruits all attached to the receptacle of the flower. These simple fruits themselves can be berries, drupes, follicles or nuts.

Observe the examples of these fruits at the front. Classify which simple fruit type develops from each individual ovary. Note that strawberry is also an **accessory fruit** because the fleshy tissue is not derived from the ovary. For strawberry list both the fruit type of each individual ovary, and the floral part that becomes fleshy.

Strawberry	
Nelumbo	
Magnolia	
Raspberry	

NOTES:

IIIi. Multiple Fruits: Some plants have fruits that develop from whole inflorescences - from flower stalks. Examples include figs, mulberries, osage oranges, and pineapple. Observe the example of pineapple at the front.

What type of fruit are the individual parts of pineapple?

Structurally how can we tell that pineapple is a multiple fruit?

NOTES:

Studying Plant Anatomy: Vegetative Structure of Vascular Plants

SECONDARY GROWTH

Introduction: To properly understand secondary growth, one must first be familiar with primary structure of the stem and the root. Specifically you should have an understanding of the organization of the primary tissues in the two stems we have studied (*Medicago* and *Coleus*) and of the *Ranunculus* root. It may be a good idea to review both "Cells and Tissues of the Plant Body", "The Root", and "The Shoot" before proceeding.

Some Important Definitions

Primary tissues: Tissues generated from the growth of an apical meristem.

Cambium: A lateral meristem consisting of a sheet of cells. Growth of these cells increases the girth of the plant organ involved.

Secondary tissues: Tissues generated from the growth of a cambium.

Vascular Cambium: A cambium that gives rise to secondary xylem to the inside, and to secondary phloem to the outside.

Periderm: A structure that consists of a cork cambium (phellogen), producing cork tissue (phellem) to the outside, and in some cases, a layer of cells to the inside called phelloderm. Periderm functions to limit dehydration and block pathogens after the epidermis is disrupted by the onset of secondary growth.

Cork (phellem, you need know only the term "cork"): Tissue, dead at maturity generated from a **cork cambium**. The cell walls of the tissue are impregnated with suberin. This waterproofs the tissue. The cork used to seal wine bottles is cork tissue harvested from a species of oak. The cell theory was first proposed by Robert Hooke in 1665 after microscopic examination of a slice of cork.

Cork Cambium: A cambial layer that functions to produce cork, and in some cases, phelloderm. In roots it is derived initially from pericycle. In stems it is first derived from the cortex. Unlike the vascular cambium, these cambial layers do not persist for the duration of the life of the plant organ. Over time, one cork cambium will be supplanted by another generated from parenchyma cells further inside.

Phelloderm: In some periderms the layer of living secondary tissue is generated by the cork cambium to the inside. We will not consider the phelloderm in the following exercise.

I. THE PRIMARY STRUCTURE OF THE *TILIA* STEM

Take the prepared slide of the cross section of a *Tilia* (basswood) stem at the end of primary growth and survey the slide with your microscope at 40x. Identify the three tissue systems. This stem differs from that of *Medicago* or *Coleu*s. In *Tilia*, the pith rays are only one cell layer wide and the primary vascular tissue appears as an almost continuous ring. As in both *Coleus* and *Medicago*, the primary xylem lies inside the phloem. As in the stems studied earlier, the ground tissue is segregated by the vascular tissue into pith and cortex. The dermal tissue consists of an epidermis. This structure will be transformed by the growth of two cambiums: the vascular cambium, which forms between the primary xylem and primary phloem, and by the cork cambium that will form in the cortex and will supplant the epidermis.

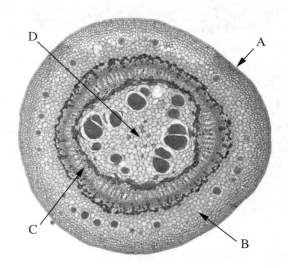

View the Figure and identify all the labels:

A = _____

B = _____

C = _____

D = _____

II. SECONDARY STRUCTURE OF THE *TILIA* STEM

Take the prepared slide of the cross sections of *Tilia* (basswood) stems: 1, 2 and 3-year sections on the same slide. Survey the section of the 1-year old stem with your microscope at 40x. Starting in the middle of the stem, identify the following: the **pith**; the **xylem tissue**, the **phloem tissue**, the **cortex**, the **periderm**, and the **epidermis**. Now switch to 400x. Carefully study the boundary of the xylem with the pith. The innermost xylem is primary xylem. The outermost is secondary xylem. The boundary between the two is subtle: the secondary xylem starts where the **xylem rays** begin. Xylem rays are areas of parenchyma of the secondary xylem running laterally through the stem. Start at the outside of the cylinder of xylem and locate a ray and follow it towards the pith. Where it ends marks the location of the primary xylem.

Study the boundary between the xylem and the phloem. This marks the location of the **vascular cambium**. Note that on either side of the vascular cambium are secondary tissues derived from the vascular cambium. Also note that the farther the tissues lie from the vascular cambium, the older the tissue. This means that **as we move outward through the xylem we move progressively from older to younger xylem**. However, **as we move outward through the phloem we move progressively from younger to older phloem**. The outermost phloem is the protophloem and is obvious because of the fiber cells in that tissue. These fibers differentiated from cells in the primary phloem that matured with the onset of secondary growth.

Outside of the phloem is the cortex. In the cortex a periderm has formed to the outside. The periderm consists of a cork cambium together with the cork tissue derived from that cambium. Note the epidermis being sloughed off the stem.

Label the figure of the one-year old cross section on the this page:

Section at the End of Three Years Growth

Switch back to 40x and survey the 3-year old *Tilia* section. The obvious changes visible here are the growth rings present in the secondary xylem, and the growth of certain rays in the phloem forming wedge-shaped regions in that tissue. The sequence of tissues outlined before are the same from the center outward: pith, primary xylem, secondary xylem, vascular cambium, secondary phloem, primary phloem, cortex, and periderm.

Switch to 400x and carefully study a growth ring of the secondary xylem. The growth increments are areas where smaller thick-walled vessel members border larger thin-walled vessel members. The smaller cells make up late summer's growth and the larger cells early spring growth. By observing this boundary you should be able to tell in which direction is the pith. You will need to make this determination for one question on the lab exam.

The rays in the xylem are continuous with those in the phloem. The enlargement of some of the phloem rays serve to relieve the tension on the phloem created by the expanding cylinder of xylem. This stress tends to create longitudinal rips in the phloem which would destroy its integrity. The expansion of these rays (they are called dilated rays) prevents these tears. The phloem outside of this ray tissue consists of bands of fibers alternating with areas containing sieve-tube members and companion cells.

Label the following figures:

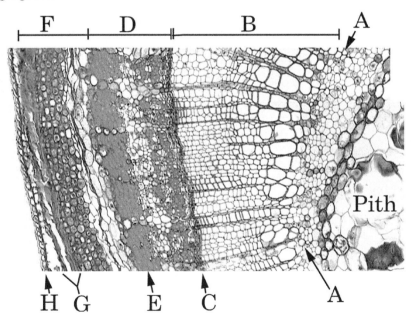

Cross Section of a One-year Old *Tilia* Stem

A = _____   E = _____

B = _____   F = _____

C = _____   G = _____

D = _____   H = _____

Cross Section of a Three-year Old *Tilia* Stem

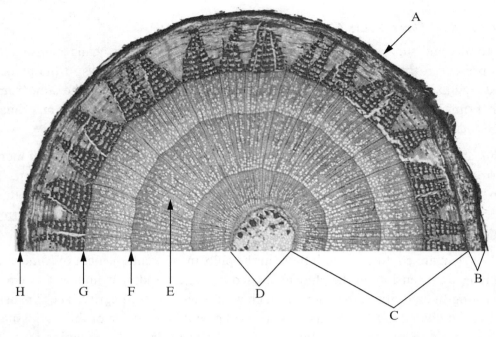

A = _____   E = _____

B = _____   F = _____

C = _____   G = _____

D = _____   H = _____

Other Notes or Sketches

Detail of the Secondary Xylem and Phloem of an Older Tilia Stem

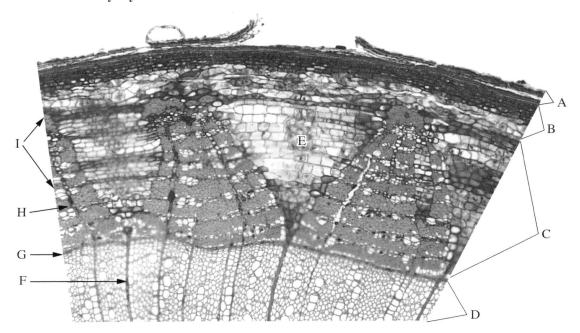

A = _____

B = _____

C = _____

D = _____

E = _____

F = _____

G = _____

H = _____

I = _____

Other Notes or Sketches

Periderm of an Older *Tilia* Stem

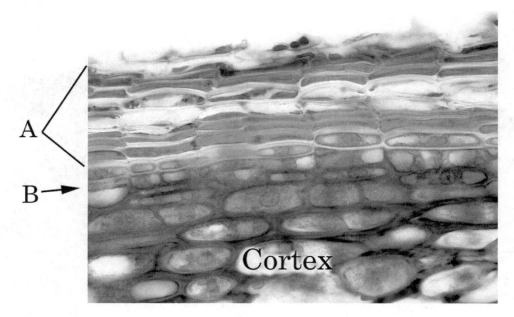

A = _____

B = _____

Other Notes or Sketches

III. FACE VIEW OF A VASCULAR CAMBIUM

The vascular cambium is a hollow cylinder of actively dividing cells. In cross sections of stems, this dimensional aspect of the structure isn't apparent. To see a the cambium as a sheet of tissue, one must observe a longitudinal section of a woody stem through the cambium. This type of section is a tangential section. The cut is made following a plane tangential to the cylinder of the cambium.

Observe the demonstration slide of the tangential section of the cambium of *Robinia* (black locust). Note that there are two types of cells each with a different orientation. One type of cell is arranged vertically and are termed **fusiform initials**. The initials of these cambium cells go on to form tracheary elements in the wood (and other cells oriented vertically in the wood), or, if incorporated into the phloem, sieve-tube members, companion cells or fibers (again all the cells oriented vertically). The other cells are arranged horizontally and are called ray initials. These go on to form the rays both in the secondary xylem and secondary phloem.

IV. LENTICILS

Lenticils are tears in the bark. Lenticils allow for the diffusion of gases to and from the living tissues in the woody stem. Observe the demonstration slide at the front of a section of a lenticil in *Sambucus*.

V. GROSS STRUCTURE OF WOODY STEMS

Woody stems are mostly secondary xylem (wood) surrounded by bark. The xylem may include heartwood and sapwood. Heartwood is darker. While its cells are all dead and the tracheary elements are nonfunctional, the tissue still provides support. The sap wood is lighter. In the sap wood the tracheary elements (tracheids and vessel members) are functional and the tissue includes living parenchyma cells. The boundary between the bark and wood marks the location of the vascular cambium. The bark itself is divided into two regions by the cork cambium: the living area inside the cork cambium is the inner bark, and the dead tissue outside is the outer bark. Evidence of earlier cork cambia can be easily discerned in some woody stems.

Take a section of oak wood from the front bench and identify the following: **pith, heart wood, sap wood, vascular cambium, xylem rays, inner bark**, and the **outer bark**. If you can't locate these all of the above, ask your TA for help.

Identify the Labels on the next page

Cross Section of a Woody Stem of Bur Oak *(Quercus macrocarpa)*

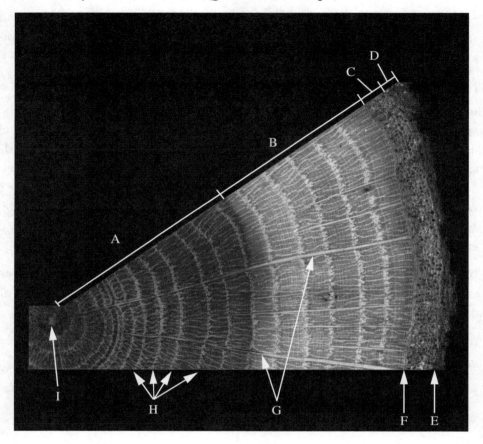

A = _____   F = _____

B = _____   G = _____

C = _____   H = _____

D = _____   I = _____

E = _____

VI. SECONDARY GROWTH IN ROOTS

The onset of secondary growth in roots is somewhat different than that in stems. This is well illustrated in your text on page 601 in figure 25-16. The pericycle plays an important role in secondary growth. It both forms the periderm and also splices together the pieces of vascular cambium at the protoxylem poles where it is discontinuous at the beginning of secondary growth. Because the first periderm is formed by the pericycle, all the tissues outside and including the endodermis are sloughed off immediately when secondary growth occurs in the root.

VIa.   ***Secondary Growth in Tilia:*** Locate the prepared slide of *Tilia* root and observe with your microscope. Identify the vascular cambium, secondary xylem, and secondary phloem. Can you identify the star-shaped mass of primary xylem inside this root? Identify the periderm, vascular cambium, primary xylem, secondary xylem, and secondary phloem.

From what primary tissue is the secondary tissue surrounding the phloem derived?

VIb.   ***The carrot:*** Carrots are roots with lots of secondary growth. At their core, they are composed of xylem tissue surrounded by phloem tissue. Make a cross section of a carrot and observe it with a dissection microscope. Note the rays which are indicative of secondary tissues. Identify the vascular cambium, primary xylem, secondary xylem, and secondary phloem.

From what primary tissue is the secondary tissue surrounding the phloem derived?

Oak wood and the inner core of carrots are both examples of xylem tissue. Why is one hard and tough while the other is crisp and delicious? Can you relate this to the different functions of a roots vs. stems?

Topics For Discussion

The wood sections studied in this topic were of *Quercus macrocarpa* (bur oak). This tree was common in the fire swept savannahs of Wisconsin. Based on what you observed today, how is this tree adpted for surviving fire?

Pinus banksiana trees are sensitive to fire. The species, however is dependant on fire. Without fire the tree would be excluded from the forests of Wisconsin. How can trees sensitive to fire persist in an environment dominated by fire?

Boreal forests are dominated by conifers (spruce and fir). In what way may tracheids be adaptive for trees subject to extremely low minimum winter temperatures?

A "girdled" tree is doomed. Exactly how does the tree die?

Some "girdled" trees will live for a season or more. Account for this delayed death.

THE SHOOT

Introduction: This is the second of two labs that focus the three higher plant organs (root, stem, leaf). There are two basic objectives for these labs: recognition of how the tissues are organized in each of the three different plant organs; and understanding how each of these organs function as a whole to ensure the plant's survival in its environment.

In most plants, stems serve to a support the leaves which act as solar collectors producing food for the plant. The tissues of the stem must also conduct water up and photosynthate down from the leaves. Stem and leaf are tightly integrated. Together they constitute the shoot system. Selective pressure in certain plant groups have resulted in modified stems and leaves that serve a number of different functions including food storage and defense. In some plants (cacti are examples), the stem is the primary photosynthetic organ and the leaves are greatly reduced. These modifications and others will be considered in the final lab of this semester.

I. COLEUS SHOOT TIP

Take a prepared slide of a longitudinal section of a *Coleus* shoot tip, and survey the slide at 40x. At the tip is the apical meristem of the shoot. Behind the meristem are a series of leaf primordia that define the nodes. Note that the internodes become increasingly elongated as one moves down the stem. Inspect the living *Coleus* plant nearest you on your bench.

Are the leaves of the plant opposite or alternate?

How are the pairs of leaves at each node oriented relative to the leaves at the nodes above and below?

Again view the slide of *Coleus*. Based on what you learned from your observations of the living plant, can you determine how many nodes are represented on your slide?

Now focus on the tissues represented on your slide. Identify the three primary meristematic tissues (**pro-toderm, ground meristem, procambium**) behind the apical meristem. Note the regions of the ground meri-stem that make up the **pith** and **cortex**. Because of the tight integration of leaf and stem, stem anatomy is a bit more complex than that of the root. Strands of procambium tissue in the stem (**leaf traces**) diverge from vascular bundles into the developing leaves (**the leaf primordia**). As they do so they leave behind an area of ground meristem tissue called a **leaf gap**. This is one of the defining characteristics of mega-phylls (remember the leaves we are considering in these anatomy exercises are fundamentally different from those found in the phylum Lycophyta! - microphylls are not even thought to be homologous to megaphylls). To gain a clearer picture of the structure of these leaf gaps study the model of the *Coleus* shoot tip at the front.

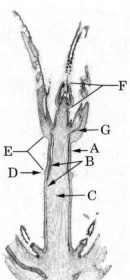

LONGITUDINAL SECTION OF COLEUS SHOOT TIP

Identify all labels: A–C refer to the three primary meristems. Other labels indicate the following: leaf primordium, node, leaf trace, and internode.

A = _____

B = _____

C = _____

D = _____

E = _____

F = _____

G = _____

Detail of a Leaf Gap

II. CROSS SECTION OF HERBACEOUS EUDICOT

This is partly a review of what we have already seen (see Lab 19 part I, Identifying Tissue and Cell Types in Cross Sections of Eudicot Stems). Here, however, we are not simply focusing on the recognition of the tissue types but on how they are organized.

Survey a slide of a cross section of a *Medicago* stem at 40x with your microscope. Note the arrangement of the **vascular bundles** dividing the **ground tissue** into **pith, cortex** and **pith rays**. This is a typical arrangement found in many eudicots. These same plants commonly have roots with tissues arranged like that of *Ranunculus* studied earlier with the vascular tissue in the center.

Structurally, would the circular arrangement of three or more vascular bundles embedded in the ground tissue be more adaptive for a stem than the arrangement of one strand of vascular tissue in the center?

Hint: What forces is the stem subject to that the root is not?

Switch to 100x and carefully study the xylem in one vascular bundle. Note that it consists of both red-stained vessel members and parenchyma cells. Also notice that the vessels become smaller towards the pith. These smallest vessels are protoxylem vessels and the larger vessels, towards the cortex, are meta-xylem vessels.

How does the direction of differentiation of the xylem in this stem differ from that of *Ranunculus* root studied earlier in Topic 20 (inside out vs. outside in)?

Now switch your view to the phloem and carefully study the tissue at 400x. Identify sieve tube members and companion cells if you can. The area of the phloem towards the cortex is the protophloem and is filled with **fibers**. Along with sieve tube members, companion cells and parenchyma cells, phloem often includes fibers as seen here.

Are there any tissues seen in this view that would typically be associated with the formation of lateral branches?

How do lateral branches originate?

Observe and identify the labels of the figure.

Details of the Medicago Stem
A = _____

B = _____

C = _____

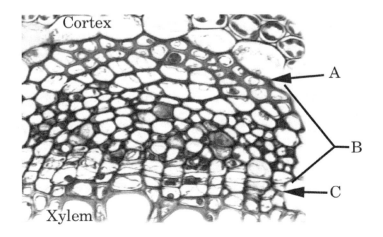

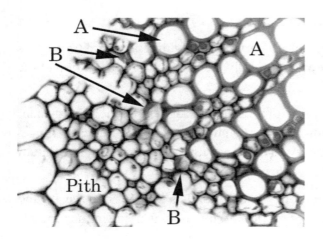

A = _____

B = _____

Study the area between the primary xylem and phloem. This is the region of the **fascicular cambium**, the vascular cambium found within the vascular bundle. These cells were derived from procambium and maintain their ability to undergo cell division. Note also the cambium layer in the pith rays connecting the fascicular cambia together to form one continuous cylinder of vascular cambium. The region of cambium formed in the pith rays that originate from ground meristem is termed **interfascicular cambium**.

Detail of *Medicago* Stem Cross Section with Cambium. Identify all labels.

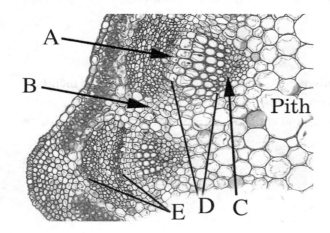

A = _____

B = _____

C = _____

D = _____

E = _____

III. THE MONOCOT STEM

Place a slide of the *Zea* (corn) stem on your microscope and survey the cross section of the stem at 40x. (Not the longitudinal section!)

How is this stem structured differently from that of *Medicago*?

Because of the arrangement of the vascular bundles the ground tissue is not clearly divided into different regions. Hence, we do not use the terms cortex or pith and we simply use the generic term "ground tissue".

Switch to 400x and closely examine a vascular bundle. First identify the xylem marked by the prominent vessels. Identify the protoxylem vessels based on their incomplete secondary walls. Now examine the phloem. Identify both sieve tube members and companion cells—this is an easy task in this tissue. Now note the sclerenchyma surrounding the bundle. This is a closed bundle with no area of cambium. Typically monocots do not have secondary growth, and, in the few cases where they do, the pattern of that growth is entirely different from that observed in Eudicots.

Figure of Vascular Bundle of Zea Stem: Identify all labels.

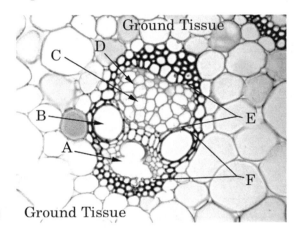

A = _____

B = _____

C = _____

D = _____

E = _____

F = _____

IV. LEAVES

Typically, leaves have determinate growth. They grow to maturity and then all growth stops—forever. As we saw with the *Coleus* shoot tip, new leaves are produced from the apical meristem of the shoot. While expanding, the leaf tissues are made up of procambium, ground meristem and protoderm. As these tissues mature, usually no apical meristems form in the mature leaf. The perenniating structures of the shoot are the buds. As covered in our first lab, a simple leaf can be distinguished from a leaflet of a compound leaf by the location of the **axillary bud**.

IVa. The Lilac (Syringa) Leaf

Survey the prepared slide of the <u>cross section</u> of *Syringa* leaf with your microscope at 40x. Note the blade and the midvein as viewed in cross section. Switch to 400x and carefully study an area of the blade away from the midvein. Note the three tissue systems. The **dermal tissue** is represented by the **upper and lower epidermis**. The **ground tissue** of the leaf is the mesophyll, and here is represented by an upper layer of elongated, vertically arranged cells (the **palisade parenchyma**) and a lower layer of horizontally arranged cells (the **spongy parenchyma**). The **vascular tissue** is restricted to the **veins**. Every cell is in close proximity to a minor vein. This is critical because the veins move water to and also move photosynthate out of each and every cell. To gain an appreciation of how pervading this network of veins is, observe the demonstration slide of a cleared leaf available at the side bench.

Things to consider while viewing the cross section of the blade

1. How might it be adaptive to have the palisade parenchyma arranged vertically?

2. Would you expect to find more stomata on the upper or on the lower epidermis?

3. How do materials move to and from the minor veins to the cells of the ground tissue (palisade parenchyma + spongy parenchyma)?

4. What structure controls the movement of materials to and from the veins?

<u>Now carefully observe the midvein:</u> Note that the xylem is on the bottom and the phloem is on top. Try to identify sieve tube members in the phloem.

Based on how the xylem and phloem are arranged in the midrib, how are these tissues arranged in the lilac stem?

Observe the figure on this page and identify all labels.

Cross Section of Syringa Leaf Blade 1,000x

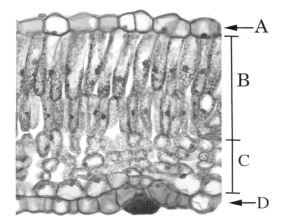

A = _____

B = _____

C = _____

D = _____

IVb. Anatomy of Leaves Adapted for Dry or Wet Environments

Syringa is a plant adapted to moist conditions. On the side bench are examples of two different leaves adapted to different moisture levels. Both of these leaves have the same basic anatomy of *Syringa* in that the mesophyll (ground tissue) include palisade and spongy layers. However each is radically different in that one is adapted to dry conditions while the other is adapted to wet conditions. Observe these slides and write down two modifications that you observe in each that allow that plant to survive in its environment. Each demonstration will include information that will guide you. You should also discuss ideas about these adaptations with your TA.

Notes for *Nerium oleander* _____

–Dry Adaptations _____

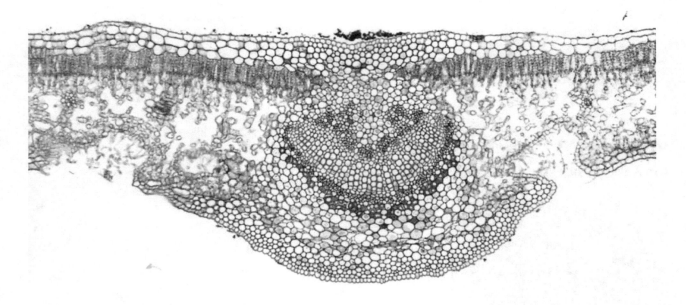

Notes for *Nymphaea oderata*—Wet Adaptations

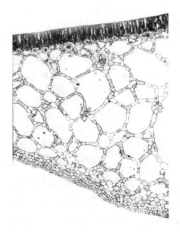

_____

_____

_____

_____

_____

_____

THE ROOT

Introduction: This is the first of two lab topics that focus on the three higher plant organs (root, stem, leaf). There are two basic objectives for these labs: recognition of how the tissues are organized in each of the three different plant organs; and understanding how each of these organs function as a whole to ensure the plant's survival in its environment.

In most plants, the root must fulfill two fundamental roles: **absorption** of water and nutrients, and **anchorage**. Further, in many plants, roots function to store food. This and other adaptations will be considered in this activity.

_____

V. GROSS MORPHOLOGY OF A YOUNG ROOT

Grass Seedling Root: Make a wet mount of a grass seedling and observe the seedling root at 40x. Identify the **root cap**. Behind the root cap is a region of growing tissue which extends back to the root hairs. This growth is due to a combination of cell division and cell elongation. Switch to 400x and carefully observe the root cap. Note cells which already seem to be partly detached. These cells would be rubbed off the root cap if the root were growing through soil. Move up the root to the region with the root hairs. Carefully study a root hair at 400x using through focusing.

Is the root hair multicellular?

What is its relationship of the epidermis to the root hairs?

VI. CROSS SECTIONS OF EUDICOT ROOT

Place a slide of a mature *Ranunculus* (buttercup) root on your microscope. Survey the entire section at 40x. Note that the organization of the tissues here are quite different from those we observed earlier in the dicot stem. The vascular tissue is in the very center of the root. The ground tissue surrounding the vascular cylinder is the **cortex**. An epidermis surrounds the entire root. The central region of vascular tissue is termed the **vascular cylinder**.

In terms of the stresses placed on a root how might this tissue arrangement be adaptive (hint: pull not push) ?

Note the innermost layer of the cortex which is stained red. This layer is the **endodermis**. The endodermis was derived from the ground meristem and is properly part of the cortex. All the tissues inside the endodermis were derived from procambium. **Xylem** fills the very middle of the vascular cylinder and its boundary is marked by ridges and valleys. The valleys are filled with phloem, and there are as many strands of phloem as there are ridges of the xylem. Note that each **phloem** strand has one enormous sieve tube member. Outside of this cylinder of xylem and phloem, located immediately below the endodermis, is a region of cells called the **pericycle.** These cells give rise to lateral roots and are also important in secondary growth.

Fill in the figure of the cross section of a mature *Ranunculus* root below.

A = _____

B = _____

C = _____

D = _____

Immature Section of a Ranunculus Root

Place the slide of an immature root on your microscope and survey the section at 40x. This cross section was made closer to the apical meristem of the root, and has immature tissues. Note that all the vessel members in the center are immature and without secondary walls.

Are there any mature vessel members in this cross section?

Where are they located?

● Based on your observations of both mature and immature roots, in which direction does differentiation progress in the procambium of this root: from the inside out or from the outside in?

The movement of materials through the immature root. Carefully observe the endodermis of the immature root. Note that these cells are not uniformly stained red as seen in the mature root. These endodermal cells have not yet deposited layers of wax and suberin along all of their walls, hence, water and solutes can freely move through the cell walls. Note however that two of the four walls seen in this cross section do have an area that stains red. This marks the presence of a suberized region of the cell walls and of the middle lamella called the **casparian strip**. This structure channels water moving inward through the root to the membranes of the cells of the endodermis. Substances must move through the living protoplast of the endodermal cells before entering into the vascular cylinder.

Cross Section of an Immature *Ranunculus* Root: Fill in the labels with the name of the appropriate tissue or structure.

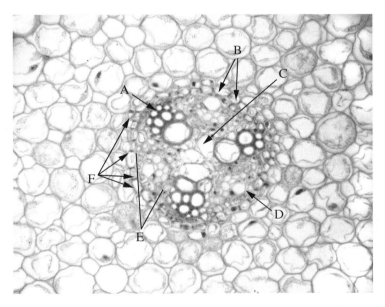

A = _____

B = _____

C = _____

D = _____

E = _____

F = _____

Another look at the mature root. Switch back to the slide of the mature root and observe the vascular cylinder. From viewing the vascular cylinder of the immature root, it should be clear that the tissue at the ridges of the xylem is protoxylem, and that in the center is metaxylem. In the cortex, note the **starch grains**. Closely observe the endodermis. The red stained area of the cell walls indicate the presence of a **suberin lamella**. This blocks all movement through the cell walls of the endodermal cells. However, materials can still pass through the endodermis via plasmodesmata (symplastically). Notice that the endodermal cells near the protoxylem vessels (at the protoxylem poles) do not have a suberin lamella. Water and minerals can still pass through the cell walls to the protoplasts of the endodermal cells at those points. These cells are called **passage cells**.

Detail of Vascular Cylinder of a Mature *Ranunculus* Root Label the indicated cells

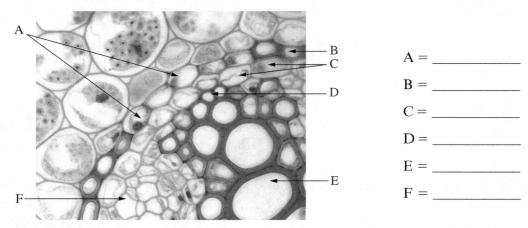

A = _____

B = _____

C = _____

D = _____

E = _____

F = _____

VII. ORIGIN OF BRANCH ROOTS

Roots don't have leaves, nodes or axillary buds. In roots branching occurs from within. Specifically branch roots form from the pericyle. Observe the demonstration series of slides on the side bench of lateral branching in *Salix* (Willow) and in *Eichhornia* (Water hyacinth).

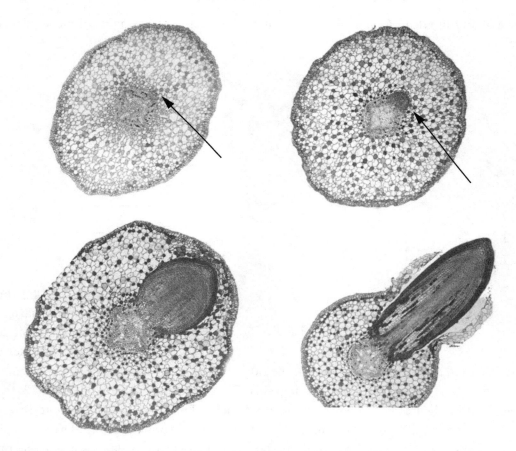

Series of Salix root cross sections illustrating the origin of lateral roots

Studying the Animal Kingdom: Phyla Porifera and Cnidaria

PRIMITIVE MULTICELLULAR SPONGES: *CELLULAR LEVEL OF ORGANIZATION*

Learning Objectives

Describe the basic forms of the sponge body.

Explain the function of the various types of cells.

Identify the skeletal structures and their composition in various sponges.

Explain the relationship of the sponges to the protista and more complex organisms.

INTRODUCTION

The Porifera are aquatic, "pore-bearing" sponges, mostly attached to surfaces in marine waters. They exhibit a division of labor among their cells and have reached the cellular level of organization. Not much more than loose aggregations of cells, the sponges are not arranged to form true tissues, organs or organ-systems. They range in size from a few millimeters to nearly 2 meters across. Although sponges are loosely two-layered as adults, there is no equivalent to the developing germ layers of higher organisms. The fertilized zygote of many sponges forms a hollow blastula with flagellated cells lining the interior surface. When the cells reverse position, the flagellated swimming larva is called an **amphiblastula.** The dividing cells develop into the specialized cells of an adult sponge.

The sponge body is generally cylindrical with many small pores or **ostia** through which water passes. Flagellated cells called collar cells or **choanocytes** line inside passages trapping food particles and producing a water current. This type of water sweeping to gather food is called **filter feeding,** a common method found in many aquatic animals. An inside chamber, **spongocoel,** opens to the outside by an upper **osculum.** Giving shape and form to the sponge is an internal skeletal array of protein (**spongin**) fibers and/or spicules of calcium carbonate (limestone) or silicon dioxide (glass). Other cells found in the sponge body include: **pinacocytes** which cover the external surface, **amebocytes** or wandering cells which help distribute food, **archaeocytes** which produce sex cells and **scleroblasts** which secrete the skeletal pieces.

The sponge phylum is subdivided into classes identified by the skeletal composition and body organization (Figure 1). Three body types are recognized in the sponges. The **ascon** form is the simplest arrangement of cells. The slender body has many ostia but only one spongocoel, one osculum and calcareous spicules. The **sycon** sponges have a tubular body but the walls are thicker and folded into channels.

From *Zoology Laboratory Workbook,* Ninth Edition by Barbara J. Schumacher and Marlin O. Cherry. Copyright © 2003 by Barbara J. Schumacher. Reprinted by permission.

FIGURE 1 Sponge Body Plans.

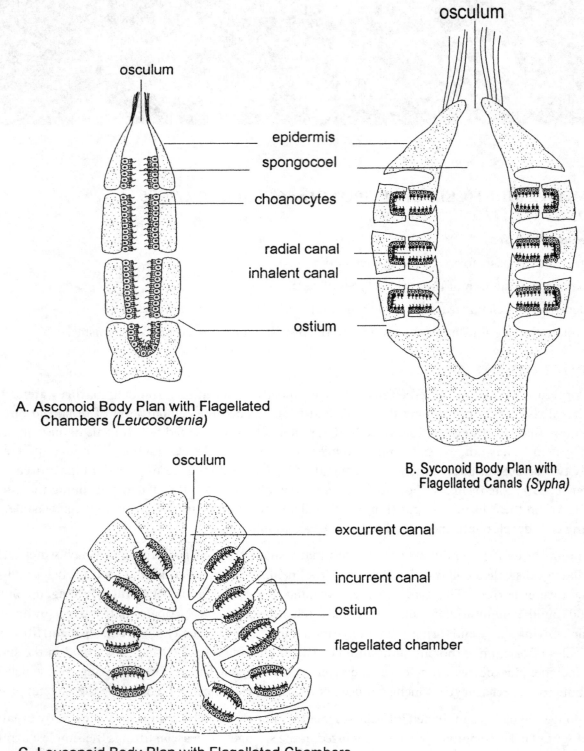

A. Asconoid Body Plan with Flagellated
   Chambers *(Leucosolenia)*

B. Syconoid Body Plan with
   Flagellated Canals *(Sypha)*

C. Leuconoid Body Plan with Flagellated Chambers
   (Euspongia)

Water entering the ostia moves into the **inhalent** or incurrent canal and passes through small openings into the **radial** canals. These channels are lined with the collar cells which move the water into the main spongocoel exiting the osculum. The most complex sponge form is the **leucon** type. Clusters of multiple chambers produce a large, nonsymmetrical mass with several oscula.

Asexual reproduction in sponges can include **budding** and **fragmentation.** They are also recognized for their high powers of **regeneration.** Experiments have shown ground up sponge bodies can reassemble cells back into fully formed organisms. Sponges possess both types of gonads resulting in a **monoecious** condition. Sperm cells are released in the water current and fertilize eggs in the chambers of other sponges. The swimming amphiblastula moves to other territories for dispersal.

CLASS CALCAREA (CALCISPONGIAE)

Example: *Scypha or Grantia*

This sponge has an internal skeleton that consists of calcium carbonate spicules. The "body" has many pores opening into one spongocoel and one osculum. Place a preserved sponge in a shallow dish of water and observe with a hand lens or dissecting scope. Identify the ostia and osculum. Obtain a prepared slide of *Scypha* (*Grantia*) and use Figure 2 to identify parts of the sponge body in both longitudinal and cross cut sections (Plate 1, 2). Look at a prepared slide of calcareous spicules (Figure 3, Plate 3). Be able to recognize the shape and composition of these skeletal units.

PLATE 1 Sycon Body Form.

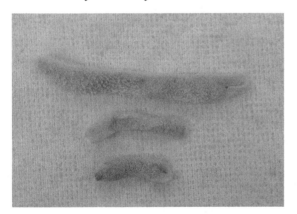

PLATE 2 Grantia (Cross-Section).

PLATE 3 Spicules (Calcareous) (Exterior).

FIGURE 2 Scypha (Grantia) Sponge.

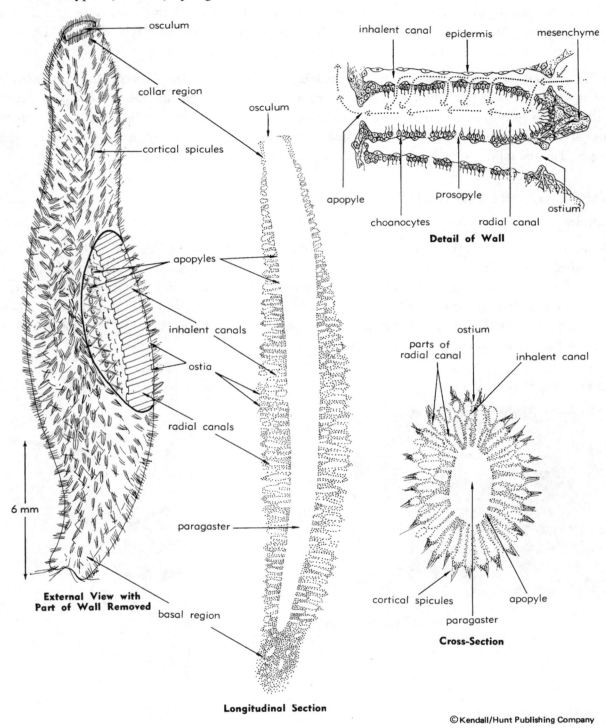

External View with Part of Wall Removed

Longitudinal Section

Detail of Wall

Cross-Section

FIGURE 3 Sponge Skeletal Structures.

A. Calcareous Spicules
   (Class Calcarea)

B. Spongin Fibers

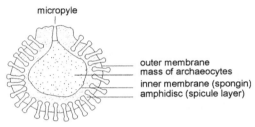

C. Cross section of a Freshwater
   Gemmule

CLASS HEXACTINELLIDA (HYALOSPONGIAE)

Example: *Euplectella* (*Venus flower basket*)

These sponges have funnel shaped bodies formed of 6 rayed siliceous spicules. The latticelike structure of the Venus's flower basket produces an exquisite form. Gently observe the preserved laboratory sample of *Euplectella.*

CLASS DEMOSPONGIAE

Example: *Spongia*

The demosponges include the majority of all living sponges. Their skeleton may contain siliceous spicules, spongin protein or combinations of both. The most popular members of this group include the "bath sponges" which are collected by sponge divers, allowed to dry and decompose, leaving their flexible skeletal remains for scrubbing purposes (Plate 4). Several bath sponges are available for display. Note the size and texture of such samples as sheep's wool and grass sponges. Obtain a prepared slide or make a wet mount of the skeleton of *Spongilla* and note the protein spongin fibers which intermesh together (Figure 3 B, Plate 5).

As a special type of survival, freshwater sponges and some marine species produce specialized packets of archeocytes called gemmules. They are surrounded by a tough hard shell embedded with spicules which allow the form to survive through adverse conditions such as drying or freezing. Observe a prepared slide of gemmules and observe the circular form with "spiked" surface (Figure 3 C, Plate 6).

PLATE 4 Common Bath Sponges.

PLATE 5 Spongin Skeleton.

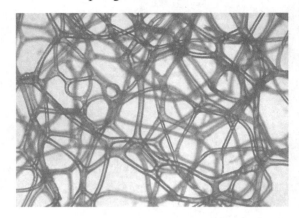

PLATE 6 Gemmule.

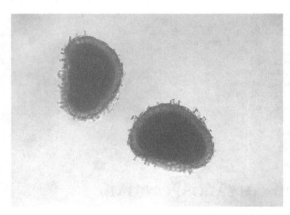

Questions

1. Why are sponges considered an "unusual" category of animals?

2. Trace the flow of water through the body of a sycon sponge.

3. Describe the basic similarity between sponge choanocytes and members of the protozoan group.

RADIATE ANIMALS: *TISSUE LEVEL OF ORGANIZATION*

Learning Objectives

Describe the function(s) of the various body tissues of the cnidaria.

Explain how digestion in cnidaria differs from that of the protista.

Explain why the division of labor among cells is necessary to increase complexity in body forms.

Explain how polymorphism relates to reproduction in the cnidaria.

Recognize specimens of representative members of the different classes of cnidaria.

INTRODUCTION
PHYLUM CNIDARIA

The cnidaria are the lowest **metazoans** (multicellular animals) with true body tissues but lack organs or systems. They include the jellyfish, anemones and corals and are identified by their **radial symmetry** and stinging cells. The **cnidocytes,** usually along tentacles, possess a triggered **nematocyst** organelle used to inject poison or attach to prey. Food is taken into the **gastrovascular cavity** through a single opening that serves as mouth and anus. The body wall consists of only two well-defined tissue layers, **epidermis** (ectoderm) and **gastrodermis** (endoderm). Between these two layers may be cellular or non-cellular jelly-like **mesoglea.**

Cnidarians illustrate **polymorphism,** different body forms within a complete life cycle. This phenomenon is illustrated both by distinct developmental forms and by differences between individual members or units of a colony. **Polyps** are generally tubular and attached stages while the **medusa** is free-swimming with tentacles dangling from an umbrella-like bell. In some cnidarians the polyp is the only stage while others may have both polyp and medusa stages, or medusa only. Most cnidarians are dioecious and aggregations of gametes arise in tissue masses, or **gonads.** In polymorphic examples the asexual budding polyps alternate with the sexually active medusa forms. A ciliated larva called the **planula** swims and settles onto surfaces to develop into new individuals or polyps.

A significant advance in the process of digestion is apparent in the cnidarians. The digestive system consists of an enclosed chamber (**gastrovascular cavity**) where **extracellular** digestion can take place. Digestive enzymes are released from **epithelionutritive** cells into the chamber and food particles are reduced to smaller particle size by chemical digestion. As a result, extracellular digestion allows the organism to utilize food particles that are larger than it can phagocytize.

CLASS HYDROZOA

Example: *Hydra*

A living freshwater hydra is a convenient animal for laboratory study. It is a solitary polyp, without a free-swimming medusa phase, and with a greatly reduced mesoglea layer. Because it does not possess a medusa form, this organism is not a "typical" cnidarian. Obtain a living *Hydra* from the jar. Add enough pond water to the culture dish to allow movement and then place the specimen in the dish. Observe the general body form with a hand lens or dissecting microscope (Figure 4, Plate 7).

FIGURE 4 Hydra (External).

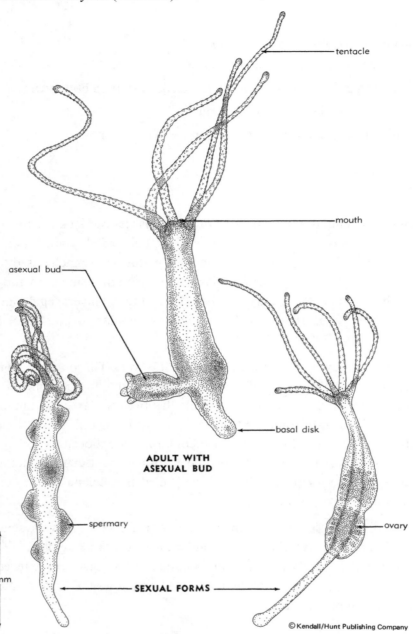

Note the normal movements of the body proper and tentacles. Observe reactions to: (1) light tapping and rotation of the dish, (2) lightly touching with a probe, (3) altering the light, (4) adding a drop of water from a culture of living *Daphnia,* and (5) adding a drop of safrain solution. Notice the discharge of the nematocysts.

Obtain different slide preparations of *Hydra.* Some specimens appear to have branches on their sides. These projections are asexual **buds.** In the sexually active *Hydra,* side bulges of **ovaries** or **spermaries** can be seen. From a cross section of *Hydra,* identify: outer epidermis, inner gastrodermis, the mesoglea between the layers and the gastrovascular cavity (**enteron**) toward the center of the ring (Figure 5, Plate 8, 9, 10).

FIGURE 5 Hydra (Internal).

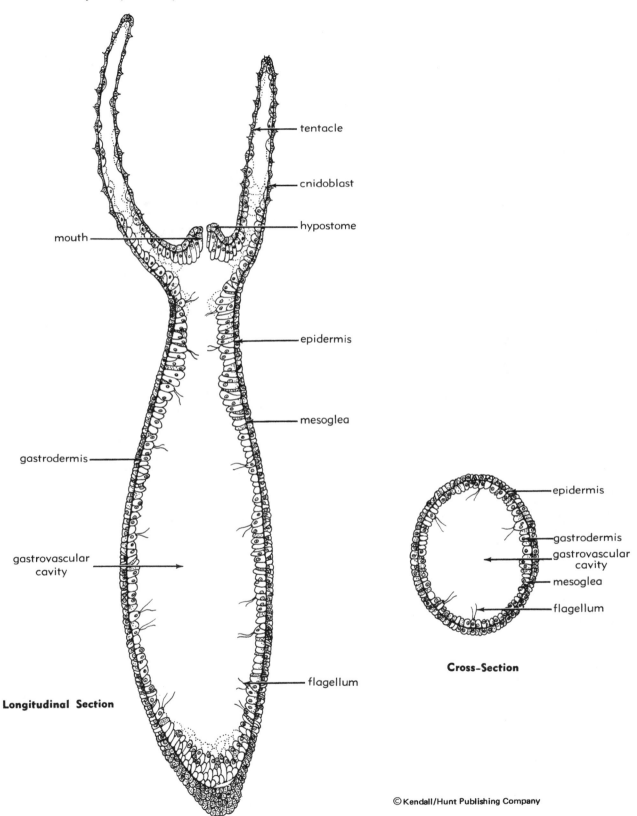

tentacle

cnidoblast

hypostome

mouth

epidermis

mesoglea

gastrodermis

gastrovascular
cavity

flagellum

Longitudinal Section

epidermis

gastrodermis

gastrovascular
cavity

mesoglea

flagellum

Cross-Section

PLATE 7 Hydra (With Bud).

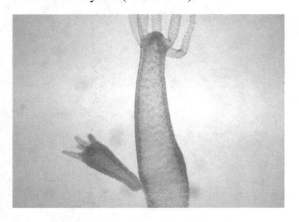

PLATE 8 Hydra (Spermary).

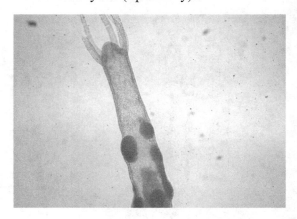

PLATE 9 Hydra (Ovary).

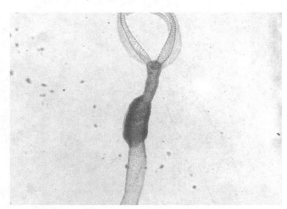

PLATE 10 Hydra (Cross Section).

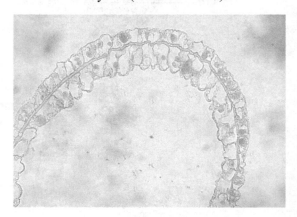

Questions

1. What type of digestion is found in *Hydra?* Compare with a sponge and a protozoan.

2. How does a *Hydra* move from place to place?

3. What special type of cell is unique to cnidarians and what purpose does this type of cell have?

4. Why is the *Hydra* an "unusual" representative of the cnidarians?

PLATE 11 Obelia (Polyp).

PLATE 12 Obelia (Medusa).

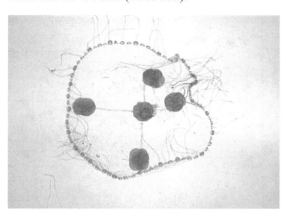

CLASS HYDROZOA

Example: *Obelia*

Polyp

The hydrozoan, *Obelia,* has an asexual stage of development in which it exists as a colony made up of two distinct types of polyps: a food gathering **hydranth** and a reproductive **gonangium.** The tentacled hydranths gather food and share it with the rest of the colony via the gastrovascular cavity. The reproductive gonangium asexually buds medusae from a club-shaped **blastostyle.** The entire colony is covered by a nonliving, chitinous **perisarc,** called the **hydrotheca** when it surrounds the hydranth and **gonotheca** when it surrounds the gonangium (Plate 11). Obtain a whole mount slide of the *Obelia* polyp. Observe each of the parts identified in Figure 6.

Medusa

Examine a whole mount slide of *Obelia* medusae. Locate the **bell** and the **manubrium** hanging from the **subumbrella** (ventral surface). Hydrozoan medusa also possess a two layered membranous ridge at the subumbrella edge called a **velum.** From the mouth food passes to the enteron and into the **radial canals** and **ring canal.** These structures are lined by gastrodermal cells in which intracellular digestion occurs. Digested food products are passed to amoebocytes that transport them to other body cells. (Figure 7, Plate 12).

Other Hydrozoans

Located on the demonstration table are preserved specimens of other hydrozoan cnidarians. Members of this class are represented by specimens such as *Gonionemus, Aequorea* and *Polyorchis.* One of the most well known yet strange jellyfish is the *Physalia* or Portuguese Man of War. It consists of a combination of many specialized polyps including reproductive, feeding, and bladders. The numerous, contractile tentacles are capable of severe stings.

CLASS SCYPHOZOA

Example: *Aurelia*

Medusa

The class Scyphozoa is represented by jellyfish in which the medusa stage is the most conspicuous stage while the polyp is minute or lacking. The amount of mesoglea in the scyphozoans is greatly increased and

FIGURE 6 Obelia Polyp Form.

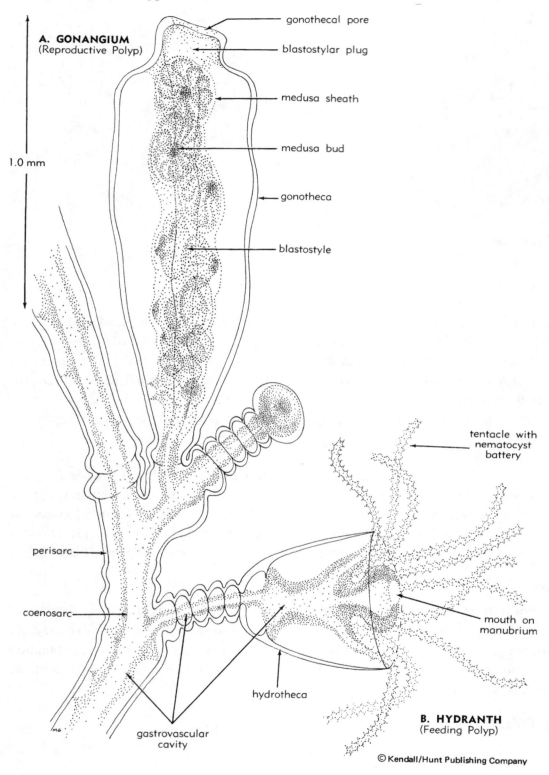

A. GONANGIUM
(Reproductive Polyp)

1.0 mm

gonothecal pore

blastostylar plug

medusa sheath

medusa bud

gonotheca

blastostyle

perisarc

coenosarc

gastrovascular
cavity

hydrotheca

tentacle with
nematocyst
battery

mouth on
manubrium

B. HYDRANTH
(Feeding Polyp)

© Kendall/Hunt Publishing Company

FIGURE 7 Obelia Medusa Form.

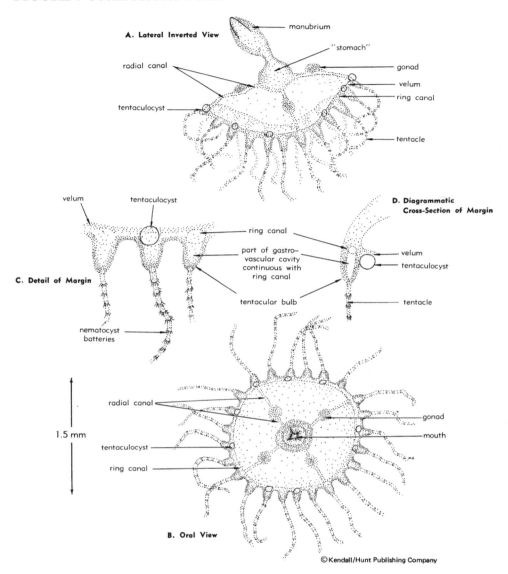

© Kendall/Hunt Publishing Company

provides buoyancy. Specialized muscle cells provide forceful contractions to propel the medusae through the water. The adult *Aurelia* is **dioecious,** being either male or female. Obtain a prepared slide of *Aurelia* medusa (Figure 8) and identify the tentacles around the bell edge, mouth, radial canal, and ring canal. Also prominent are: distinctive flaps of the mouth area or **oral arms** used for food gathering, the **gastric pouches** surrounding the mouth area, and the **gonads** closely attached. In scyphozoan medusa, the velum membrane is not present, but the edge of the umbrella is marked by specialized sensory organs for light (**ocelli**) detection and **statocysts** for orientation.

Polyp

Aurelia fertilization produces a ciliated **planula** larva. When it settles, the larva develops into an attached hydra-like polyp called **scyphistoma.** Transverse fission of the top surface in **strobilization** forms a stacked "saucer" appearance of the **strobila.** As each extension or **ephyra** develops, the strobila elongates. When mature, each ephyra breaks off and swims away to develop into a mature medusa (Plate 13 A–D).

FIGURE 8 Aurelia (External).

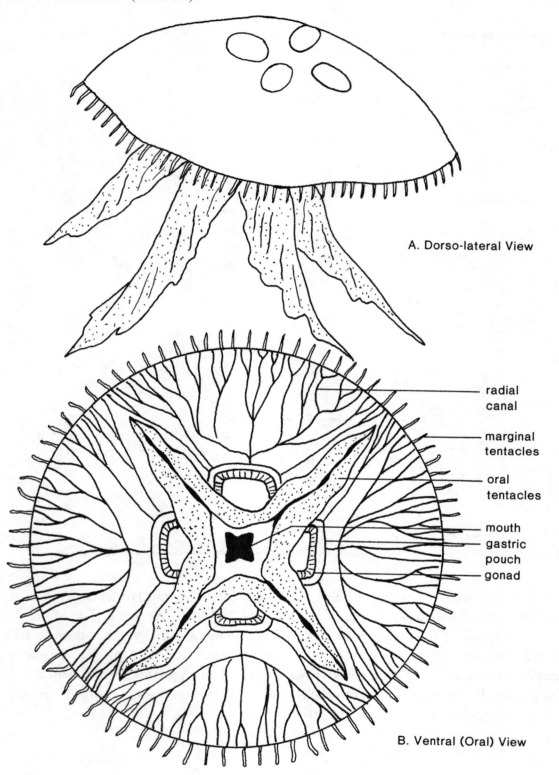

A. Dorso-lateral View

radial
canal

marginal
tentacles

oral
tentacles

mouth
gastric
pouch
gonad

B. Ventral (Oral) View

PLATE 13A Planula Larva.

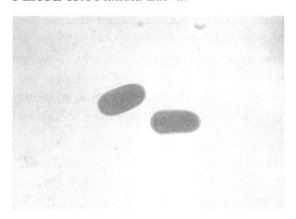

PLATE 13B Scyphistoma.

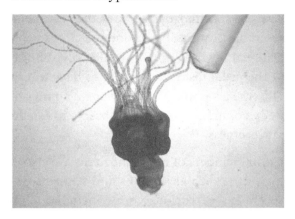

PLATE 13C Strobila.

PLATE 13D Ephyra.

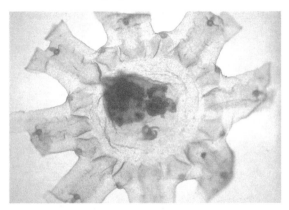

Questions

1. Describe some distinct differences between hydrozoan and scyphozoan cnidarians.

2. What are some of the advantages of having a large mesoglea-filled medusa form?

3. Identify: the site of asexual reproduction in *Aurelia*; the site of sexual reproduction in *Aurelia*.

CLASS ANTHOZOA

The class Anthozoa includes the sea anemones and corals. The anthozoans are marine animals that resemble flowers. Therefore, the Greek root word *anthos,* which means flower, is used as a prefix for the class name. They are all sessile polyps but their shapes, colors and forms vary tremendously. Attached at the **pedal disk,** sea anemones have a tubular body with multiple tentacles surrounding an **oral disc.** Observe a sample of *Metridium,* a common sea anemone of both U.S. coasts. Note the thick, muscular body column. If available, observe live anemones in a marine tank to better appreciate their beauty and gracefulness.

Coral members of the anthozoans are also extremely varied. **Hard** or **soft** corals may produce soft swaying forms such as sea fans and sea pens or lay down the calcium "skeleton" of the coral reef. Observe specimens of various corals including stony corals such as brain coral. *Astrangia* forms encrusting colonies along North Atlantic coastlines while staghorn or antler coral forms a recognizable branched colony. *Gorgonia,* or the sea fan, produces colorful shapes in tropical waters. *Renilla* (sea pansy) has a flattened, leaf-like appearance supported by a short attachment stalk. (Figure 9, Plate 14A–C)

PLATE 14A Sea Anemone.

PLATE 14B Coral Types.

PLATE 14C Sea Fan.

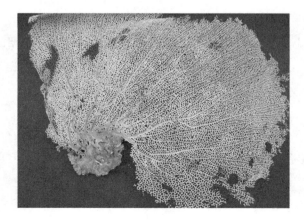

FIGURE 9 Sea Anemone.

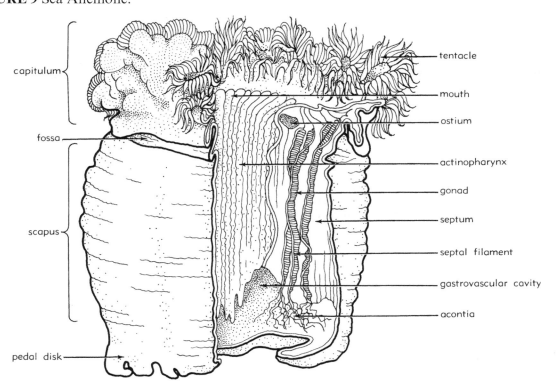

capitulum

fossa

scapus

pedal disk

tentacle

mouth

ostium

actinopharynx

gonad

septum

septal filament

gastrovascular cavity

acontia

LATERAL VIEW (One Quarter Cut Away)

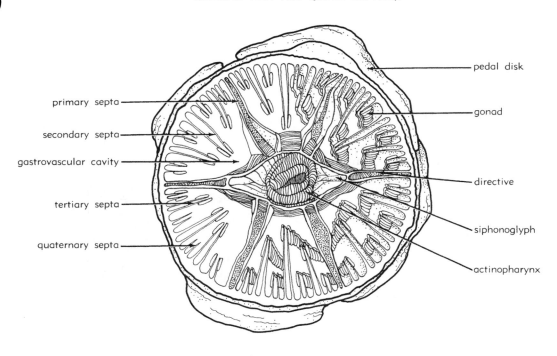

primary septa

secondary septa

gastrovascular cavity

tertiary septa

quaternary septa

pedal disk

gonad

directive

siphonoglyph

actinopharynx

Cross-Section   © Kendall/Hunt Publishing Company

Questions

1. Do anthozoans exhibit polymorphism? Explain.

2. Describe the ecological impact of coral reefs.

3. Do the anthozoans have any economic importance to man? Explain.

PHYLUM CTENOPHORA: THE COMB JELLIES
INTRODUCTION

The ctenophores are close relatives of the cnidarians. They possess the characteristic primary radial symmetry in adult form. However, they do not form colonies, do not exhibit polymorphism and lack the stinging nematocysts. They are recognized by comb-shaped plates of cilia, or **ctenes,** used for locomotion. They are monoecious and members may or may not possess tentacles. Many ctenophores, such as *Pleurobrachia* and *Mnemiopsis,* are capable of **bioluminescence** creating spectacular light shows on dark nights. If possible, observe prepared or jarred specimen of various ctenophores (Figure 10).

FIGURE 10 Ctenophore.

Questions

1. Compare and contrast the cnidaria and ctenophora.

2. How is bioluminescence possible?

3. If stinging cells are absent in ctenophores, how do the tentacled forms gather food?

Studying the Animal Kingdom: Phyla Platyhelminthes and Nematoda

FLATWORMS: ORGAN LEVEL OF DEVELOPMENT

Learning Objectives

Describe the acoelomate body plan and how this restricts the size of the flatworms.

Explain how the organ level of development in flatworms differs from the tissue level of development in the cnidaria.

Identify representative specimens of the different classes of flatworms.

Describe the life cycle of important human parasites within this group.

INTRODUCTION

As the name implies, the platyhelminthes are dorso-ventrally flattened and are called flatworms. They are the first animals to demonstrate the organ level of development, bilateral symmetry, and the simplest form of body organization built largely of mesoderm. They are **triploblastic** or have three germ layers (endoderm, ectoderm, and mesoderm) from which other tissues are derived. The addition of mesoderm allows for the development of a greater variety of tissues which, in turn, makes organ development possible. In the cnidarians, the space between the epidermis and gastrodermis was a noncellular mesoglea. This limited the number of tissues and the variety of body forms. In the flatworms, the mesoderm is a cellular material, including muscle and parenchyma. The development of specialized organs allows for division of labor and a more complicated body form.

Flatworms do not have a **coelom** (body cavity). Body cavities are those spaces lying between the inner surface of the body wall and the outer surface of the digestive tract. Lacking a body cavity, the flatworms are said to be **acoelomate.** The cells that pack the body space between the digestive tract and the body wall do not allow the free flow of fluids, gases, nutrients, and waste to circulate as well as a fluid-filled cavity does. To compensate, the body is flat to allow for adequate diffusion of these materials.

The body of flatworms exhibits bilateral symmetry which allows only one plane of division creating two equal halves. Other phyla, like the cnidarians have more than one plane of division and have radial symmetry. Bilateral symmetry seems to be better adapted for an active life. Because one end of the animal tends to be the forward moving end all the time, locating sensory structures at this end allows the animal to detect stimuli rapidly. This location also promotes brain formation and, in later phyla, head formation. Muscular movement, controlled by a more centralized nervous system, becomes more efficient.

The flatworms include free-living and parasitic forms. When a digestive tract is present, there is a single opening that leads into a "blind" pouch. Most are monoecious and have well developed reproductive systems. They range from a few millimeters in length to extremely large specimens in the parasites. Complicated life cycles have been identified in the parasitic flukes and tapeworms.

CLASS TURBELLARIA

Example: *Dugesia (Euplanaria)*

Turbellarians are mostly free-living flatworms and inhabit slow-moving streams and ponds. *Dugesia* is a common laboratory sample and exhibits the typical characteristics of the flatworms. Observe living *Dugesia* with a hand lens or dissecting microscope. Although mostly bottom dwellers, note the muscular contractions which allow a swimming motion. Cilia along the bottom (ventral) surface makes the *Dugesia* move gracefully over surfaces. Look for the distinctive **eyespots** used for light detection and the extended **auricles** along the side. Chemically, these detect the presence of food and tactile receptors interpret information about surface texture. Feed the living *Dugesia* and watch the extension of the muscular **pharynx** as it draws food into the digestive tract (Figure 1, Plate 1).

Obtain a prepared slide of *Dugesia.* A whole mount slide usually has plain and stained flatworms. Observe the body structures such as eyespots, auricles and pharynx (Figures 2, 3, Plates 2, 3). Focus on the darkened digestive tract. The pharynx leads into multiple branches of the complex **intestine.** A prepared slide of the cross sectioned *Dugesia* allows observance of internal tissue. Starting at the top surface, identify the ectoderm (epidermis), mesoderm (parenchyma) and ciliated lower epidermis. Three branches of the digestive tract are visible: the larger middle ring is the single, muscular pharynx surrounded by its sheathing chamber and the two smaller rings are the intestinal branches. The endoderm (gastrodermis) lines the digestive branches.

Although not normally visible, the planarians posses a very primitive brain, the **anterior ganglion** and a **"ladder"** branching of longitudinal nerves. The production of waste material by the multiple layered cells necessitates an excretory process. Many sac-like structures bear tufted cilia (**flame cells**). These produce a cleansing flow of fluid past cells, into tubules exiting through minute pores to the outside surface.

CLASS TREMATODA

Example: *Clonorchis (Opisthorchis)*

All trematodes are parasites and most are internal parasites. *Clonorchis* occurs in human populations of the Far East and Southeast Asia and inhabits the bile duct. Obtain a prepared side of *Clonorchis* and compare its size to that of the planaria (Figure 4, Plate 4). Note the **anterior sucker** which surrounds the mouth. It serves as a hold-fast organ to attach the fluke to its host. Posterior to the anterior sucker is the **ventral sucker (acetabulum)** located in the middle portion of the body. Like the anterior sucker, the posterior sucker is a hold-fast structure. The **mouth** is surrounded by the anterior sucker. Posterior to the mouth is a muscular **pharynx** which pumps food into the digestive tract. A short **esophagus** extends

FIGURE 1 Dugesia (Planaria).

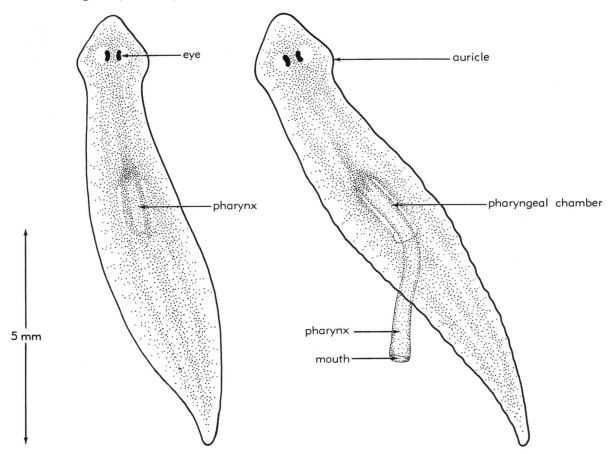

External View

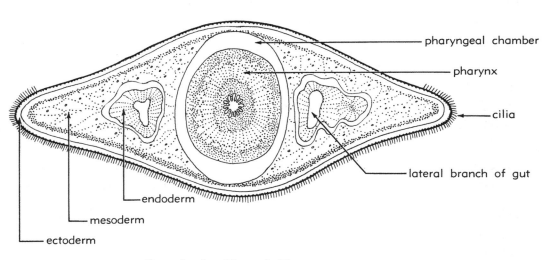

Cross-Section Through Pharynx

PLATE 1 Dugesia (Whole Mount).

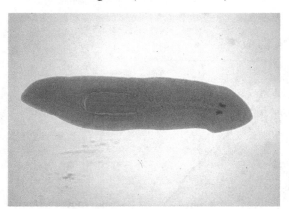

PLATE 2 Dugesia (Digestive).

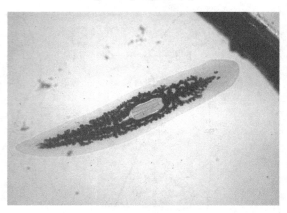

PLATE 3 Dugesia (Cross Section).

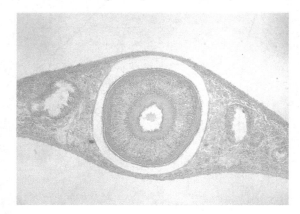

posteriorly from the pharynx and divides into Y-shaped branches called the **intestinal crura.** Like planaria, *Clonorchis* has no anus: therefore, it has an incomplete digestive tract.

At the posterior end is the **excretory pore** near the excretory **bladder** which it drains. Flukes have both male and female reproductive structures; therefore, they are monoecious but usually cross-fertilize. The female system consists of a single lobed **ovary** about one third of the distance from the posterior end. A large **seminal receptacle** for sperm storage is just posterior to the ovary. Two **yolk (vitelline) ducts** connect the lateral **yolk glands (vitellaria).** The **oviduct** then extends anteriorly and enlarges into a highly coiled **uterus** filled with encysted eggs. Eggs leave the uterus through the **genital pore** near the ventral sucker. The male system consist of two large lobed **testes** in the posterior region of the fluke. The **vasa efferentia** connects each testis to the single **vas deferens.** The vas deferens extend anteriorly to the male **genital pore** near the ventral sucker.

The life cycle of *Clonorchis sinensis* is shown in Figure 5. Fertilized eggs are defecated into the water and must be ingested by a specific snail species. These larval **miracidia** enter the snail's tissue and transform into a bagged **sporocyst** filled with **rediae** (sing; redia). As the rediae move into the snail's liver, they transform again into tailed **cercaria.** These tadpole-looking forms emerge into the water and burrow into the muscle of specific fish species as encysted **metacercaria.** Ingestion of raw or poorly cooked fish with metacercariae releases them into the human digestive tract. Heavy infestations of liver flukes generally results in cirrhosis of the liver and even death.

FIGURE 2 Dugesia (Planaria) Internal.

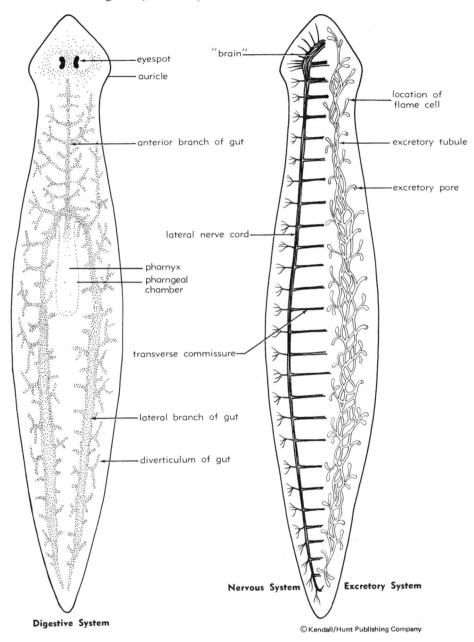

Digestive System

Nervous System   Excretory System

© Kendall/Hunt Publishing Company

Other Flukes

If prepared slides or jarred specimens are available, observe other flukes. *Fasciola hepatica,* a very large liver fluke, is found in deer, goats and sheep. It's anatomy is very similar to *Clonorchis,* except for its impressive size. Observe as many body structures as possible using a dissecting microscope. Blood flukes or **schistosomes** invade the veins of human intestines and urinary bladders. A significant difference of *Schistosoma* is that they are dioecious. Larger, broader males have a groove along their length in which the slender female is embraced.

FIGURE 3 Lateral View of Pharnyx (Dugesia).

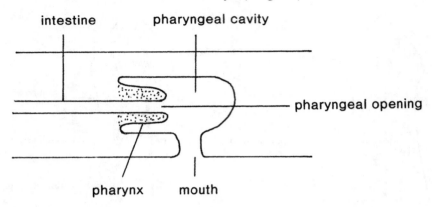

PLATE 4 Clonorchis.

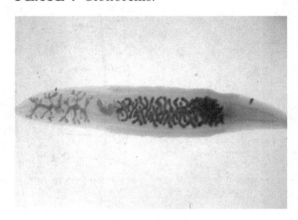

CLASS CESTODA

Example: *Taenia*

Tapeworms are all internal parasites and usually need two hosts to complete their life cycle. The adult parasitizes the digestive tract, but the immature forms may be found in various organs. Since the adult form lacks a digestive tract, the parasite simply absorbs the products of the host's digestive process. By locating in the digestive tract, the eggs made by the adult tapeworm have an exit to the environment where additional hosts may be infected. By locating in organs that may be eaten, the immature form gains access to a new host.

On prepared slides of *Taenia,* various sections of the tapeworm have been mounted. Progressing from the smallest to largest sections, identify the various parts. Since the adult tapeworm lacks a digestive system, most of the internal structures belong to the reproductive system. The body consist of three regions: the anterior **scolex,** a short **neck** region and a long ribbon-like body known as the **strobila** (Figure 6). The scolex is a permanent structure and has **hooks (rostellum) and/or suckers** at the anterior end. These structures serve as holdfast organs. At the posterior end of the scolex is the **neck** which continuously buds new sections or **proglottids.** Like the fluke, the tapeworm is monoecious so each proglottid contains male and female reproductive organs.

FIGURE 4 Clonorchis (Liver Fluke).

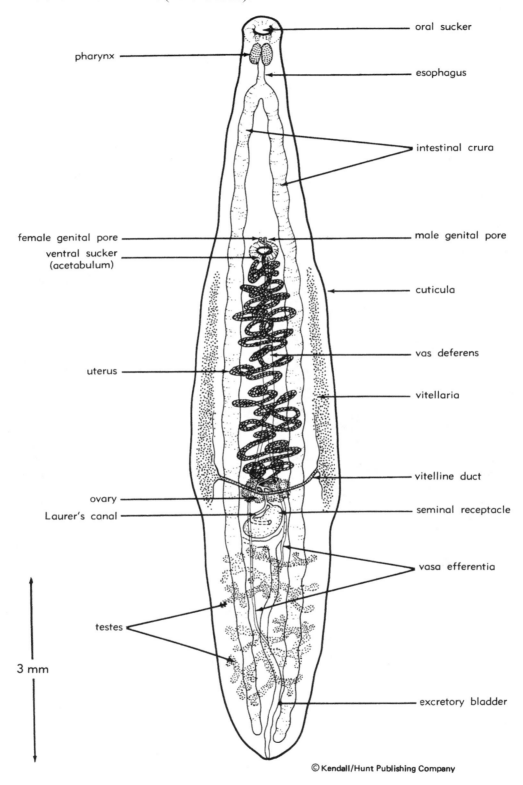

© Kendall/Hunt Publishing Company

FIGURE 5 Clonorchis Life Cycle.

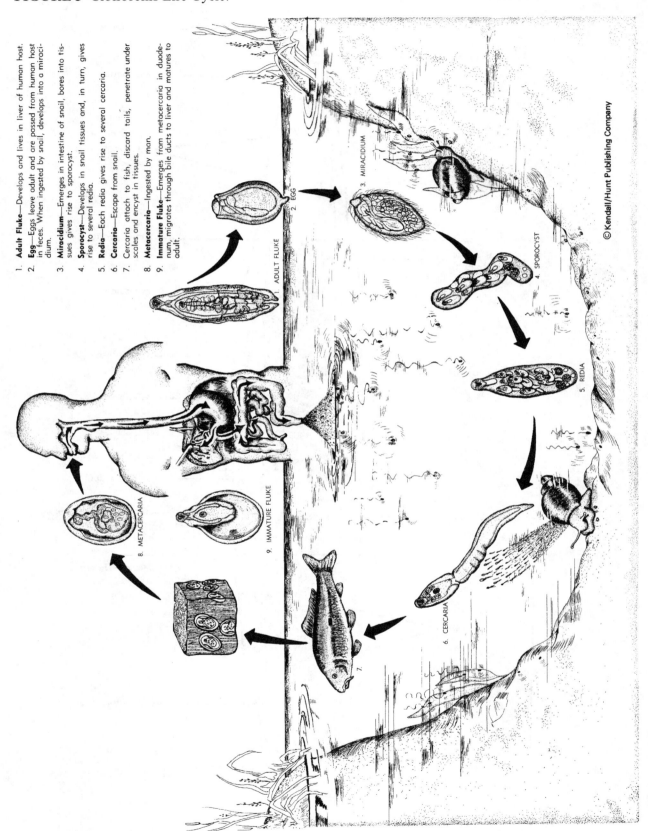

1. **Adult Fluke**—Develops and lives in liver of human host.
2. **Egg**—Eggs leave adult and are passed from human host in feces. When ingested by snail, develops into a miracidium.
3. **Miracidium**—Emerges in intestine of snail, bores into tissues gives rise to sporocyst.
4. **Sporocyst**—Develops in snail tissues and, in turn, gives rise to several redia.
5. **Redia**—Each redia gives rise to several cercaria.
6. **Cercaria**—Escape from snail.
7. Cercaria attach to fish, discard tails, penetrate under scales and encyst in tissues.
8. **Metacercaria**—Ingested by man.
9. **Immature Fluke**—Emerges from metacercaria in duodenum, migrates through bile ducts to liver and matures to adult.

1. ADULT FLUKE
2. EGG
3. MIRACIDIUM
4. SPOROCYST
5. REDIA
6. CERCARIA
7.
8. METACERCARIA
9. IMMATURE FLUKE

PLATE 5A Scolex.

PLATE 5B Gravid Proglottid.

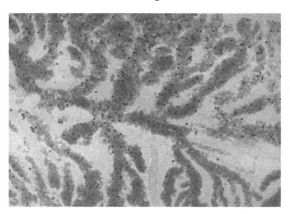

The most anterior proglottids are known as **immature** proglottids whose reproductive structures are just developing. Consequently, these structures are ill-defined and hard to see. Along the lateral edge of each proglottid is a lightly stained line, an **excretory canal.** The excretory canals connect at the posterior end of each proglottid with a **transverse canal.** These empty waste from the body at the posterior end of the tapeworm. One or two large swellings or **genital pores** are located at the midpoint of each proglottid. From the genital pore extends two ducts. The more anterior duct is the **sperm duct** of vas deferens and the posterior duct is the **vagina.** The vas deferens branches into smaller ducts or **vasa efferentia** that collect sperm from the many small testes that fill the proglottid. The vagina extends posteriorly toward the **ovary** where it then connects with a short oviduct, the yolk glands, and a long slender **uterus.**

Gravid proglottids are filled with encysted eggs. The other reproductive organs are usually difficult to observe because of the numerous eggs. These fertilized eggs develop into six-hooked larva known as **onchospheres.** Passed out in the feces and ingested into the digestive tract of the **intermediate host,** these larva attach to the wall of the intestine. Obtain a prepared slide of the larva **cysticerus** (Figure 7, Plate 5 A&B) or **bladderworm.** This stage in the beef and pork tapeworms produces "measly" meat. Note the distinct head region but usually the neck and body are reduced in size and invaginated. When the intermediate host or parts of the intermediate host are eaten, the bladderworm evaginates and attaches to the intestinal wall of the **definitive** host.

Other Tapeworms

Located on the demonstration table are examples of other tapeworms. Observe jarred specimens of *Taenia saginata* (beef) or *Taenia solium* (pork) and note the extensive length of an extracted adult. If slides are available, note the similar structures in *Dipylidium caninum,* a tapeworm commonly infesting dogs.

FIGURE 6 Tapeworm (General Structures).

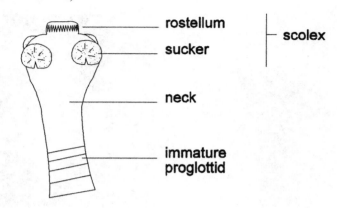

A. Scolex and Neck Region

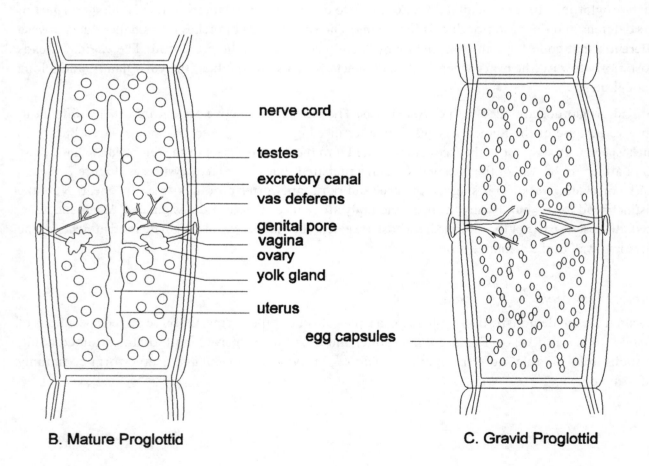

B. Mature Proglottid   C. Gravid Proglottid

FIGURE 7 Cysticercus (bladderworm).

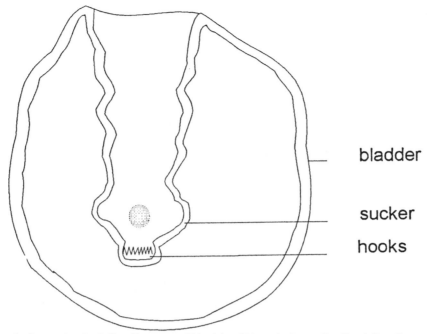

bladder

sucker

hooks

A. Invaginated Cysticercus (Bladder Worm), Longitudinal Section.

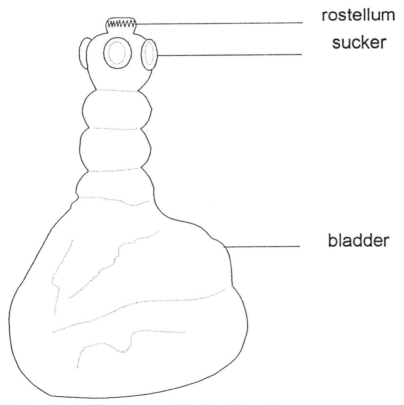

rostellum

sucker

bladder

B. Evaginated Cysticercus (Bladder Worm).

Questions

1. What structures differentiate the anterior end of planaria? Of liver flukes? Of tapeworms?

2. What type of symmetry do the flatworms possess?

3. How is locomotion accomplished in planarian worms?

4. How is food ingested in the planarians?

5. How do the flukes and tapeworms get food without digestive organs?

6. What is a flame cell? Explain its function.

7. How could infections by flukes and tapeworms be halted?

8. What is the advantage of being a monoecious parasite?

PHYLUM NEMATODA
ROUNDWORMS: COMPLETE DIGESTIVE TRACT
Learning Objectives

Describe how the organ-system level of complexity in this group differs from that of the flatworms.

Explain the importance of the pseudocoelom and its advantages over the acoelomate body plan.

Identify representative specimens of the different classes of nematodes.

Recognize the life cycles of parasitic nematodes and identify preventive techniques of noninfestation.

INTRODUCTION

The phylum Nematoda is the first group of animals to be studied that has a **complete** digestive tract, a tube within a tube body plan. This allows for food to be broken down and absorbed as it moves along the tract from the anterior to the posterior end of the animal. A muscular **pharynx** at the anterior end ingests the food and the undigested portion passes out a second opening known as the **anus.** Therefore, food does not have to be regurgitated to be eliminated. Nematodes also have a type of body cavity known as a **pseudocoelom.** This fluid-filled chamber allows nutrients absorbed from the gut tract to freely circulate from front to rear within the worms. No cell membranes act as barriers to slow the movement of nutrients as in the flatworms. Consequently, the pseudocoelom serves as a simple type of circulatory system. It also functions as a hydrostatic skeleton to stiffen the body. Being **bilaterally symmetrical,** the body is streamlined for movement. These two factors allow for efficient use of the muscular system.

Nematodes are chiefly free-living and microscopic. They range is size from microscopic to a meter in length. They occupy a great number and variety of habitats in soil, mud, and fresh or salt water. Some soil nematodes, however, are plant parasites and cause millions of dollars in agricultural damage annually. Many animals, including humans, are parasitized by nematodes such as pinworms, *Trichinella,* numerous filarial worms, and several species of hookworms.

Example: *Ascaris lumbricoides*

Ascaris is not a typical nematoda. It is used because of its large size and availability. Most of its pseudocoel is filled with a tremendously elongated coiled reproductive system. The female is capable of producing 200,000 eggs daily. Although infection by Ascaris is **direct,** the path of the larva through the body is complicated. Ingested eggs are swallowed, hatch in the intestine but then burrow through the wall and enter the bloodstream. At the lungs, the immature worms break out into the alveoli, are coughed up, swallowed and finally reach the intestine for anchorage.

Obtain a preserved specimen of *Ascaris* and place in a dissecting pan. Most nematodes are dioecious and there are usually external differences sufficient to distinguish them apart. Look for the obvious size differences between males and females. The male is smaller and the tail end is conspicuously curled with a pair of copulatory **spicules.** On the larger female find the **head** with three **lips** surrounding the **mouth.** The largest lip is dorsal, and the other two are ventro-lateral (Figure 8). The anal end is somewhat larger than the anterior end.

Internal Anatomy

After selecting a female worm, dissect the worm as follows. Start to slit the animal open along the its length beginning at the head region. The internal organs are very soft so barely insert the point of a dissecting needle just under the body covering and pull the tip along. The body wall will split open easily. Using large dissecting pins, pin the body wall on either side exposing the organs. Identify the muscular **pharynx** and the ribbon-like **intestine** which extends to the anal opening. Look for a "wishbone" like structure about one third down the length. The short **vagina** is attached to the **genital pore** and splits into two large **uterus** branches. The eggs produced in the terminal **ovaries** move through the coiled, tubular **oviducts.** As they pass into the uterus, stored sperm fertilize them.

Examine a prepared slide of male and female *Ascaris* worms. In the cross section of the male, note the flattened **intestine** and the numerous sperm ducts or **vas deferens.** Identify the **cuticle** covered body wall, the **muscle** layer beneath and the clear cavity of the **pseudocoelom.** In the cross section of the female, similar structures will be evident. However, the two large **uteri** filled with eggs are usually the most predominant structures. The smaller circular structures are the **oviducts** and **ovaries** (Figure 9, Plate 6 A, B).

Example: *Trichinella*

Examine a prepared slide of *Trichinella* and observe under low-power magnification. Look for the encysted worms rounded up within the muscle fibers. Humans become infected with *Trichinella* by eating contaminated and poorly cooked pig meat (Figure 10, Plate 7).

Example: *Enterobius vermicularis*

The pinworm or seatworm is a cosmopolitan parasite of the human intestinal tract with a direct life cycle including **self-contamination.** Upon ingestion or inhalation of embryonated eggs, development continues in the intestinal tract where attachment, mating and egg production occurs. Females migrate to the anal opening at night to deposit the eggs on the skin surface. This accounts for the easy recontamination by children as they scratch the irritated perianal area. Inhalation of eggs on bedclothes or sheets is also a common infestation mode by adults attending infected children. Observe a prepared slide of *Enterobius* and find the light pointed tail region which gives it one of its common names (Figure 10).

Example: *Necator americanus*

These hookworms possess a slightly curved anterior end whose mouth area has cutting plates. Attached to the intestinal wall they feed on a blood meal resulting in symptoms of anemic conditions. Infestation by juveniles in the soil usually occurs through skin surfaces, primarily the soles of the foot. Migration through the body is similar to the *Ascaris* trek. Look at a prepared specimen of *Necator* or *Ancylostoma.* Compare it to *Enterobius* and be able to distinguish between them (Figure 10).

Example: *Dirofilaria*

The filarial worms invade fluid tubes such as blood and lymph vessels. The common *Dirofilaria* infects dogs through the bite of mosquitoes. The larval forms are identified in prepared slides of blood tissue and can be seen as slender worm forms between the blood cells. Matured worms live in the heart and lung arteries resulting in enlarged organs and eventually heart and lung failure (Figure 10, Plate 8).

FIGURE 8 Ascaris Anatomy.

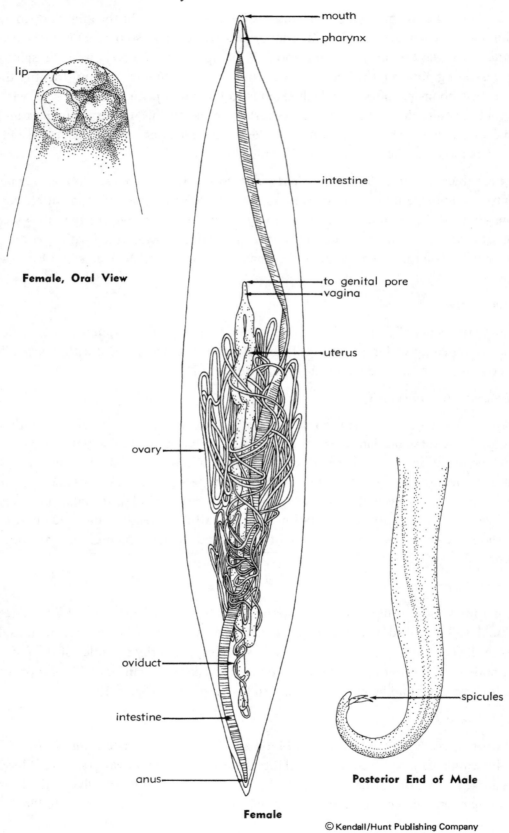

lip

Female, Oral View

mouth
pharynx

intestine

to genital pore
vagina

uterus

ovary

oviduct

intestine

anus

Female

spicules

Posterior End of Male

© Kendall/Hunt Publishing Company

FIGURE 9 Ascaris (Female and Male).

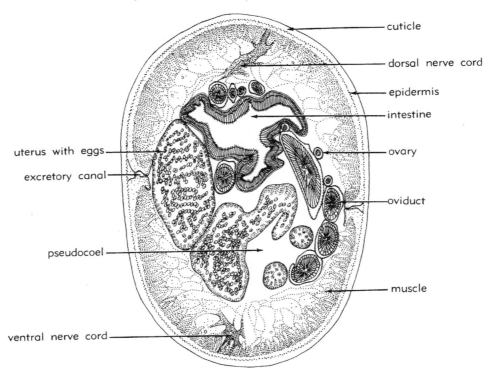

Female

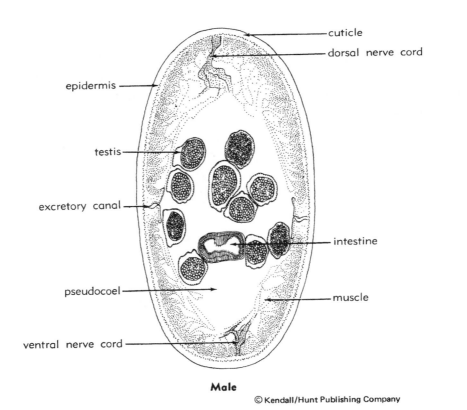

Male

PLATE 6A Ascaris (Male).   **PLATE 6B** Ascaris (Female).

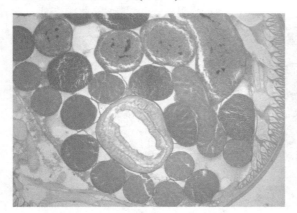   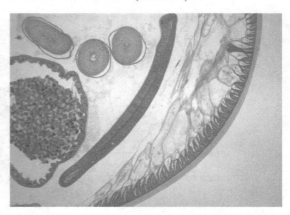

PLATE 7 Trichinella.

Other Nematodes

If available, watch living vinegar eels (*Turbatrix*) move about in the vinegar solution. They thrive on the bacteria and yeast there. Another significant nematode is *Caenorhabditis*. It has become a classic research animal in studies of genetics and developmental control. Because of its limited number of body cells, scientists have been able to map the source of all body cells in the adult worm and are studying the genetic control of differentiation.

FIGURE 10 Parasitic Roundworms.

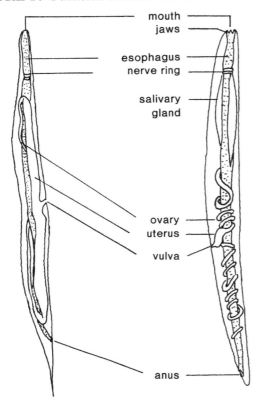

A. Pinworm, Female

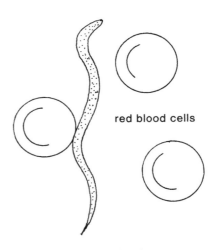

C. Microfilaria of
Dirofilaria immitis

B. Hookworm, Female

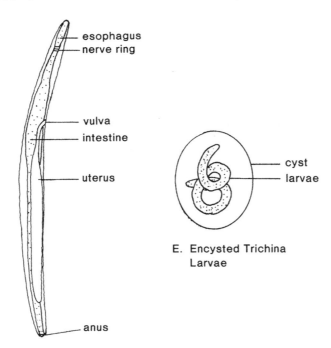

D. Trichina Worm, Larvae

E. Encysted Trichina
Larvae

PLATE 8 Dirofilaria

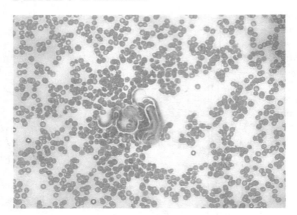

Questions

1. What type of symmetry is found in the nematodes?

2. What type of diet do the different types of nematodes have?

3. Give a comparison between flatworms and nematodes.

4. Explain the term dioecious. Is this an advancement over monoecious animals?

5. What are the advantages of a pseudocoelom?

6. How might a heavy infestation of *Trichinella* affect muscle function?

7. What is the most practical medical test/exam to determine the presence of intestinal nematodes?

8. Identify safety precautions to prevent infection by specific parasitic nematodes.

Studying the Animal Kingdom: Phyla Mollusca and Annelida

SOFT BODY FORM

Learning Objectives

Describe the general body plan of the mollusc.

Describe the structure and function of the mollusc shell.

Recognize representatives of selected molluscan classes.

INTRODUCTION

The phylum Mollusca is a well-defined but extremely diverse group of animals. The body is bilaterally symmetrical, unsegmented, and soft with a **mantle** covering. The mantle's epithelium secretes a calcareous **shell** in most members of the group. Most also have a ventral foot, a dorsal, visceral **body mass**, true **coelom**, and gills. Some like the oysters have no head or foot. Adapting to different means of nutrition and methods of locomotion has created a wide variety of body forms among the molluscs. The squid specializes in speed and predation, the clam for filtering detritus, and the snail for gliding and grazing. They have adapted to marine, aquatic, and terrestrial habitats. Molluscs have well developed excretory, respiratory and reproductive systems. The **open** circulatory system pumps blood through vessels but also allows blood to bathe tissues directly in **sinus** areas. The coelom cavity is reduced to an area just around the heart. Most molluscs are dioecious and they reproduce sexually.

CLASS BIVALVIA

Example: *Unio or Venus*

Bivalves are molluscs such as mussels, clams, oysters, and scallops that have two shells or **valves**. The shells are tightly closed by well-developed **adductor muscles**. Bivalves feed by filtering food from an internally controlled water flow.

Shell Features

Look at the cleaned shell of a bivalve clam. Determine the body orientation by looking for the swollen area known as the umbo. The umbo is the oldest part of the valve and is on the **anterior**, **dorsal** side of

FIGURE 1 Bivalve Shell.

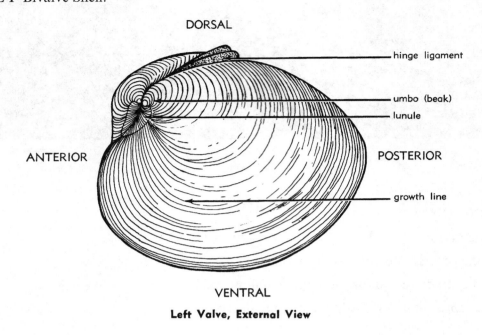

DORSAL

hinge ligament

umbo (beak)

lunule

ANTERIOR

POSTERIOR

growth line

VENTRAL

Left Valve, External View

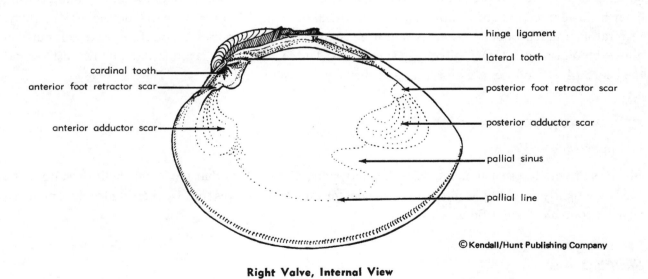

hinge ligament

lateral tooth

cardinal tooth

anterior foot retractor scar

posterior foot retractor scar

anterior adductor scar

posterior adductor scar

pallial sinus

pallial line

© Kendall/Hunt Publishing Company

Right Valve, Internal View

the clam. The extended end away from the umbo is the **posterior** end. The open edge of the clam opposite the hinge is the **ventral** margin. Note the **lines of growth** on the valves. The wide band of each represents a season's growth (Figure 1, Plate 1A–B).

PLATE 1A Bivalve Shell (Exterior).   **PLATE 1B** Bivalve Shell (Interior).

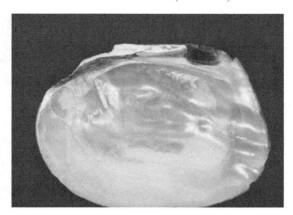

By placing the proper shell "palmed" in the hand, place the thumb on the umbo with the short, anterior end pointed upwards. This should place the ventral margin at the fingertips. The correctly palmed shell can then be designated **right** or **left** valve. (If the shell does not "set" properly, switch it to the other hand and try again.) Look at the various layers of the shell surface. The outer is the thin, horny layer or **periostracum**. The next layer is the **prismatic** layer and may be seen when the periostracum is scraped or chipped away. The innermost layer is the **mother-of-pearl** or **nacre**.

On the internal shell surface, just at the hinge are the longitudinal **lateral teeth**. These teeth prevent movement of the two valves in either the dorsal or ventral direction. Just inside and below the umbo are the pointed **cardinal teeth**. These teeth prevent any anterior or posterior movement in the valves (Figure 1). Powerful **adductor muscle scars** can be seen on either side of the hinge and the pallial line follows the curve of the ventral shell edge. This line is the scarred surface on the valve where the mantle edge attaches.

Obtain a fresh-water clam and place it in the dissecting pan. The shell is held apart by a wooden peg. Remove the peg and insert a heavy bodied knife on either side of the umbo, severing the muscles. Pry open the shell and scrape the soft body parts into the bottom half. Examine the ragged edges of the torn **adductor** muscles and note their thickness. Focus on the soft body and fold back the soft, thin membrane or mantle. Two openings at the **posterior** end of the clam include the **incurrent siphon** and the **excurrent siphon**. Water enters through the incurrent siphon, circulates around the mantle cavity, then through the gills. Filtered water is discharged through the excurrent siphon. Food particles and oxygen are removed as water circulates through the clam (Figure 2A–B, Plate 2).

Remove the exposed mantle half, carefully cutting it with scissors. Below it lies the **visceral mass**, and **foot**. A pair of striped **gills** lie on each side of the visceral mass. Locate the slit-like **mouth** opening between the foot and the anterior adductor muscle. It is bordered by a pair of small triangular **labial** (oral) **palps** that aid in passing food-laden mucus strings into the slit-like *mouth*.

Locate a cavity directly in front of the posterior adductor muscle and just below the hinge. The wall of this cavity is a thin and semitransparent membrane. Cut through the membrane to expose the **pericardial cavity** in clams, which contains the **heart**. It is a triangular structure wrapped around a portion of the digestive tract called the **rectum**. Below and lateral to the pericardial cavity is a pair of dark organs, **kidneys**, which drain the coelom.

Next, remove the body from its remaining valve. Trim away the mantle and gill tissue with scissors. Using a sharp scalpel, dissect the foot through the visceral mass by **cutting along the ventral margin**

FIGURE 2 Bivalve (Internal).

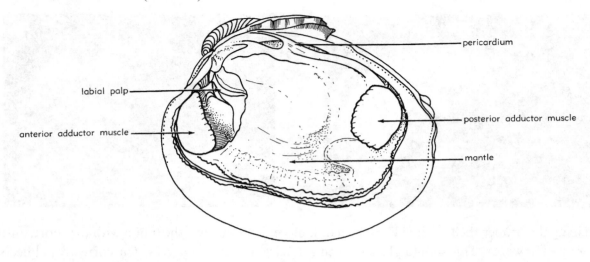

Left Valve Removed, Mantle Intact

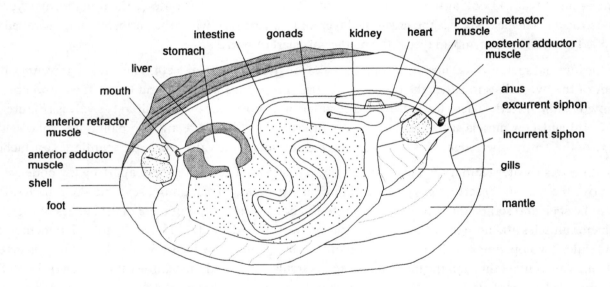

toward the dorsal margin. Dissect only the visceral mass. Locate the mouth and the short **esophagus**. The esophagus leads to the **stomach**, an enlarged sac. Masses of greenish tissue on either side of the stomach are **digestive glands**. Leading from the stomach is the twisted tube or **intestine** that is embedded in a mass of yellowish tissue called the **gonads** (either testes or ovaries). The rectum is a portion of the intestine that enters the coelom. Trace the rectum over the posterior adductor muscle where it ends at the **anus**.

CLASS POLYPLACOPHORA

Chitons

Chitons are members of the class Polyplacophora (Figure 3, Plate 3). They have an elliptical body with a shell containing a mid-dorsal row of eight broad plates. Chitons have a large, flat foot surrounded by a row of gills. Pick up and observe the ventral and dorsal surfaces of preserved chitons. Identify the overlapping plates that give the class its name.

PLATE 2 Clam Dissection.

A. Clam, intact organs.

B. Clam, dissected organs.

FIGURE 3 Classes of Molluscs.

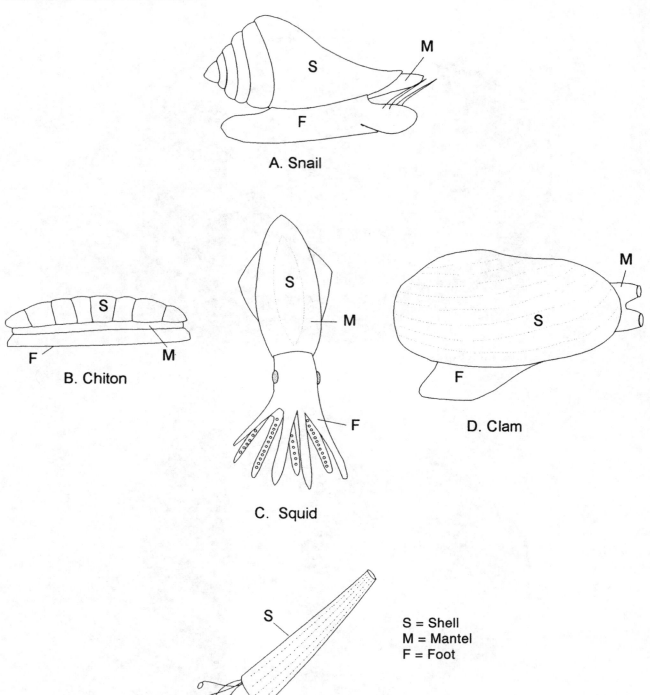

A. Snail

B. Chiton

C. Squid

D. Clam

E. Tooth Shell

S = Shell
M = Mantel
F = Foot

PLATE 3 Chiton.

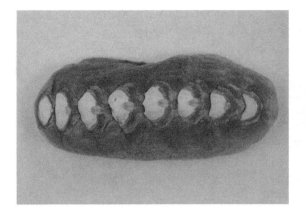

PLATE 4 Tooth Shells.

CLASS SCAPHOPODA

Tooth or Tusk Shells

These elongated, single shelled mollusks inhabit both shallow and deep oceans. With an opening at both ends of the tubular shell they can burrow headfirst into sand or mud while the siphons circulate water through the shelled surface exposed to the water. Look at some tusk shells and note the delicate shell formations (Figure 3, Plate 4).

CLASS MONOPLACOPHORA

Members of this group were identified by fossil shells and thought to be extinct. However, living specimens of *Neopilina* are now recognized as living examples of this class. The shell is a circular, cap-like dome that protects the soft body. If *Neopilina* specimens are available, note their similarities to limpet shells.

CLASS GASTROPODA

Snails and slugs are universal molluscs that are members of the class Gastropoda. The shell is usually coiled, reduced, or absent. There is a distinct head, commonly with eyes and tentacles. Most gastropods also have a rasping structure or **radula** within the mouth area. Look at a prepared slide of a snail radula and note the overlapping ridges which allow the snail to scape algae or other debris from rock surfaces (Plate 5). Locomotion is by a large, flat muscular foot. Many of the marine gastropod snails are known to produce harmful toxins that are injected into prey. Nudibranchs and sea slugs are naked gastropods, often with extended projections on their dorsal surface for gas exchange. Some brightly colored species warn of the dangerous poisons contained in those projections. Observe various specimens of this class including: conch shells, slugs, land snails (*Helix*), nudibranchs, cone shells, abalone and limpets (Figure 3, Plate 6).

PLATE 5 Snail Radula.

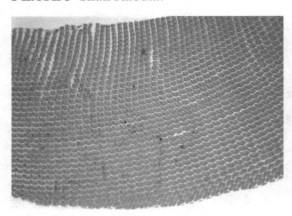

PLATE 6 Gastropod Shells.

FIGURE 4 Squid Anatomy.

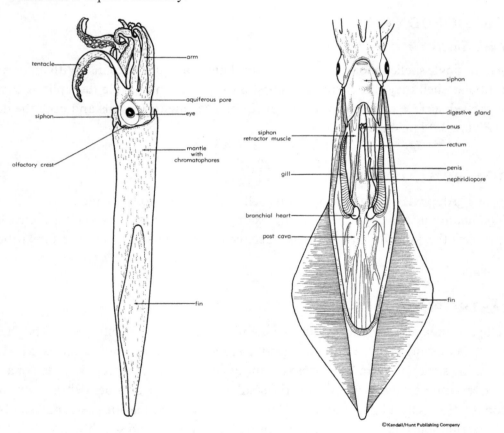

CLASS CEPHALOPODA

The class Cephalopoda is represented by the squid and octopus. The squid has a large head which is pointed with fins and conspicuous eyes. The mouth is surrounded by ten tentacles. The octopus has a rounded head with large eyes, but no fins. The mouth is surrounded by eight arms. The chambered *Nautilus*, an exquisite deep-diving cephalopod has a coiled shell separated into compartments. When new sections are added, the soft body moves forward into the completed chamber and the remaining fluid-filled chambers are pressurized for buoyancy. Look at the various specimens of the cephalopod group on display including: the cuttlefish (*Sepia*), octopus, squid and *Nautilus* (Figure 4, Plate 7A–B).

PLATE 7A Squid.

PLATE 7B Nautilus.

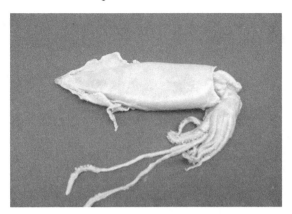

Questions

1. Why are the clam adductor muscles larger than the retractors?

2. Describe the various modes of ingestion found in molluscs.

3. What are the different layers of the molluscan shell and their function?

4. Of what economic importance are the molluscan members?

5. How is the octopus eye like a human eye?

SEGMENTED WORMS: *TRUE COELOMATES*

Learning Objectives

Explain how metamerism in the annelids has contributed to the complexity of the annelid body form.

Explain how regionalization of the internal organs has contributed to the complexity of the annelid body plan.

Identify representative specimens of the various annelid classes.

INTRODUCTION

The phylum Annelida is a group of segmented worms that include earthworms, marine bristle worms, and leeches. The annelids have a true coelom (**eucoelom**) and a **closed** circulatory system. It is the simplest of the eucoelomate animals, yet, it possesses important organ systems characteristic of all higher forms.

Perhaps the most outstanding feature that separates annelids from other worms is their **segmentation.** This repetition of body parts is also called **metamerism** and is internal as well as external. Metamerism is important because it has provided a redundant supply of organs, each of which is specialized for new and different functions. In short, metamerism provided a means by which a greater variety of organizational plans was possible. In spite of the possibilities for advanced forms, only in arthropods and chordates do we see the extent of body forms that metamerism makes possible. The compartmentalization of the body cavity by septa makes the hydrostatic skeleton more efficient. Aiding this is the presence of two muscle groups (longitudinal and circular) in the body wall and the fusion of the nervous system into one unit.

CLASS OLIGOCHAETA

Example: *Lumbricus terrestris*

Oligochaete earthworms live in moist, humus soil. They are nocturnal animals and sometimes spoken of as "night crawlers." Other oligochaete characteristics are internal and external segmentation, a reduced head or head lacking, paired **setae** per somite, monoecious, and a **clitellum.**

EXTERNAL ANATOMY

Obtain a preserved earthworm, hand lens, and a dissecting pan. Note the obvious separation of the body into **segments** or **metameres.** The mouth and **prostomium** are located at the anterior end, and the anus at the posterior end of the animal. The **clitellum** is a girdle-like structure about one-fourth of the worm's length from the anterior end. It is important in the formation of mucus during copulation and the secretion of the **cocoon** or **egg case.** The dorsal surface is darker than the ventral surface due to a faint dark **dorsal blood vessel** that runs down its length. With the earthworm between your fingers, pull gently. Chitinous bristles, or **setae** should be felt along the ventral and lateral surfaces. Each segment has four pairs of these setae. Use a hand lens or dissecting scope to see them in a dried worm specimen. As the skin contracts, they are more clearly visible along the surface (Figure 5, Plate 8).

Locate paired openings on the fourteenth and fifteenth somites. The openings on the fourteenth pair are **oviduct** openings and the fifteenth pair are **sperm duct openings** (vas deferens openings).

FIGURE 5 Earthworm (External).

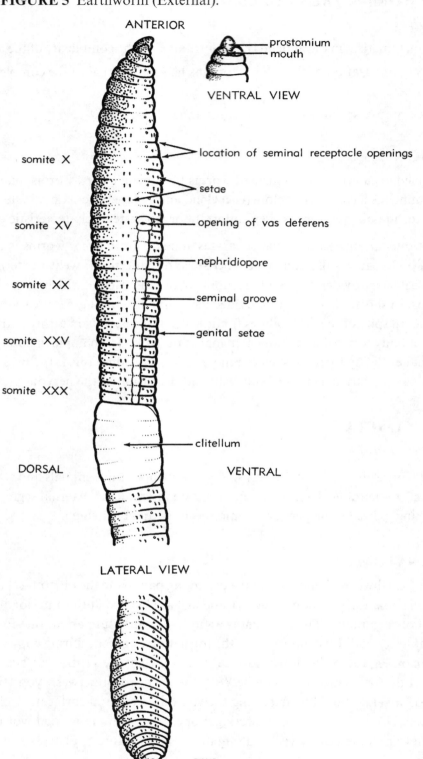

PLATE 8 Eathworm (External).

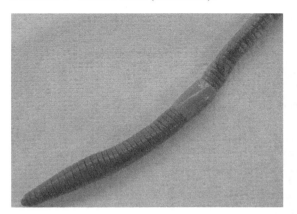

PLATE 9 Earthworm (Internal).

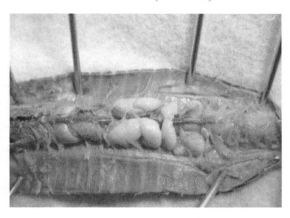

PLATE 10 Earthworm (Cross Section).

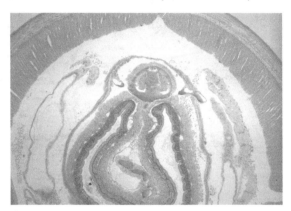

INTERNAL ANATOMY

Position the earthworm dorsal side up in the pan. To stabilize the specimen, place a pin through the middle of the animal about **midway** between the anterior and posterior end. Stretch the worm slightly and pin the worm through the **fourth** somite from the anterior end. With a scalpel, make a small slit in the dorsal surface **posterior** to the clitellum and continue the incision along the mid-dorsal line to the fourth somite. Avoid damaging the internal structures by taking care to just cut through the body wall. Beginning at the point of the **initial** incision, open the worm by pinning back the cut edges of the body wall for approximately one inch. Then cut the internal partitions, or septa, along each side of the cut with a **probe.** Do not push the pins in vertically but slant them outwards so they will not obscure the view. Continue pinning the wall for about one inch and cutting the partitions to the fourth somite. Remove the pin at the fourth somite. Cut and pin the rest of the body wall as previously described. Do not damage the brain which is just under the body wall of the **first** segment (Figure 6, Plate 9).

The digestive system includes a **mouth** and muscular **pharynx.** A relatively long (Segments 6-13) **esophagus** connects the pharynx to the storage **crop** and muscular **gizzard.** Using the probe, push on both bag-like structures and determine their texture. Trace the **intestine** down to the **anus.** Make a transverse cut of the intestine. Cut a section about three mm. in length. Wash the matter out with water. Notice the dorsal fold within the intestine called the **typhlosole** (Figure 7, Plate 10).

FIGURE 6 Earthworm (Internal).

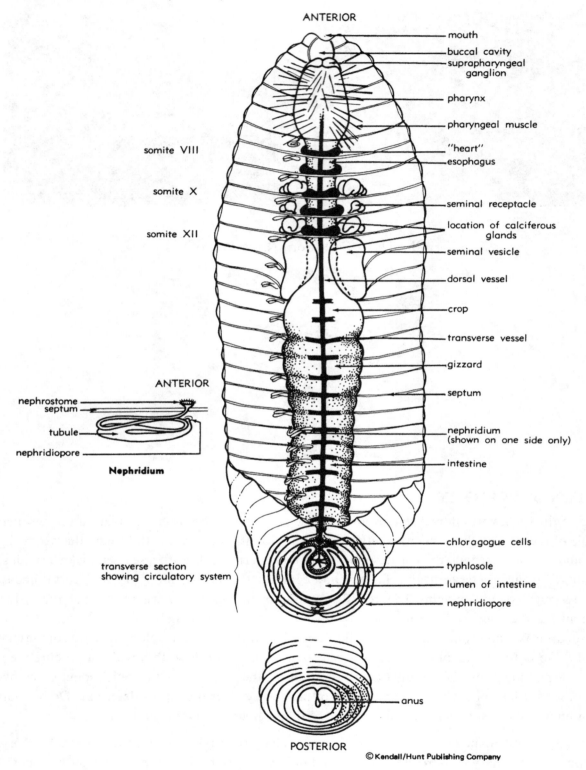

ANTERIOR

— mouth
— buccal cavity
— suprapharyngeal ganglion
— pharynx
— pharyngeal muscle
— "heart"
— esophagus
— seminal receptacle
— location of calciferous glands
— seminal vesicle
— dorsal vessel
— crop
— transverse vessel
— gizzard
— septum
— nephridium (shown on one side only)
— intestine
— chloragogue cells
— typhlosole
— lumen of intestine
— nephridiopore

somite VIII

somite X

somite XII

ANTERIOR

nephrostome
septum
tubule
nephridiopore

Nephridium

transverse section showing circulatory system

anus

POSTERIOR

© Kendall/Hunt Publishing Company

LUMBRICUS

FIGURE 7 Earthworm (Cross Section).

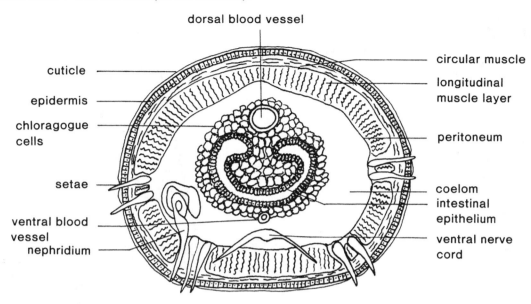

At the head region, search for five dark tubular hearts or aortic arches. These branch off the dorsal blood vessel and cover the esophagus area. Blood flow moves forward in the dorsal blood vessel into the arches and continues back to the tail in the ventral blood vessel (The ventral blood vessel is obscured by the various organs).

Look closely in any coelomic compartment—except the first three and last one. Each kidney or **nephridia** is a coiled, whitish-colored tube. With the aid of a hand lens or dissecting microscope, find a number of nephridia. Notice that each compartment contains a pair of nephridia.

The earthworm is monoecious, having both male and female reproductive organs. Identify three pairs of large, light **seminal vesicles** and two pairs of "beady" **seminal receptacles.** Other sexual parts are too small to identify in this dissection. To find the **circumpharyngeal** and **suprapharyngeal** ganglia (brain), carefully dissect away the pharynx. The small clusters of ganglia can be seen just on either side of the pharynx.

Obtain a prepared slide of the cross-section of **Lumbricus.** Identify: the **cuticle** covering the external **epidermis** and the **muscle layer** beneath it. Regions of both **circular** and **longitudinal** muscles can be differentiated. Find the rounded **intestine** with the surrounding **chloragogue** cells. Pay close attention to the infolded wall of the intestine or **typhlosole.** The **dorsal** and **ventral blood vessels** can be located attached above and below the intestine. Near the muscle wall of the ventral surface is also the circular tube or **ventral nerve cord.** Beneath the muscle layer and surrounding the intestine lies the true body cavity or coelom. If the cross cut section has cut through them, the paired ventral and lateral **setae** may be visible. Paired **nephridia** will appear as coiled tubules on either side of the intestine in the coelomic cavity.

CLASS POLYCHAETA

Bristleworms

On the demonstration table are specimens of polychaetes such as the sandworm, *Nereis* or *Neanthes.* Its segmented body has lateral extensions with bristles or **parapodia.** *Arenicola,* or lugworm, burrows in muddy soil along coastal areas. *Chaetopterus,* the parchment tube worm, produces a U-shaped tube. Specialized parapodia flatten out into fan-like wings directing a water current through the tube and across the body (Figure 8, Plate 11).

FIGURE 8 Representative Annelid Classes.

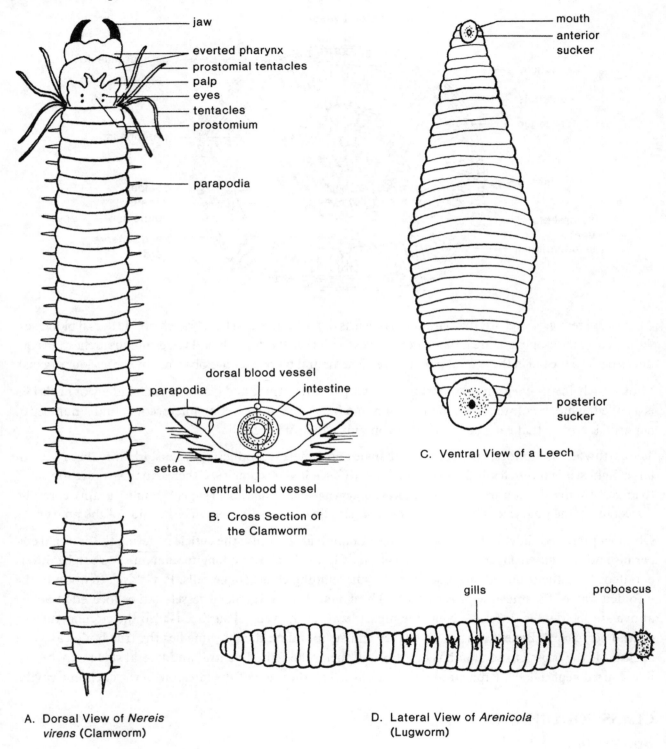

jaw
everted pharynx
prostomial tentacles
palp
eyes
tentacles
prostomium

parapodia

parapodia

dorsal blood vessel
intestine

setae

ventral blood vessel

B. Cross Section of
the Clamworm

mouth
anterior
sucker

posterior
sucker

C. Ventral View of a Leech

gills   proboscus

A. Dorsal View of *Nereis
virens* (Clamworm)

D. Lateral View of *Arenicola*
(Lugworm)

PLATE 11 Polychaete.

PLATE 12 Leech.

CLASS HIRUDINEA

Leeches

Leeches are annelids that have anterior and posterior suckers for locomotion and attachment. They may be aquatic, marine, or terrestrial. They resemble most annelids except that they lack the characteristic setae or parapodia and have a set number of segments. Observe specimens of the medicinal leech, *Hirudo medicinalis*. The blood sucking parasites have an anticoagulant in their mouth secretions and they are still used today to remove accumulated blood from bruised skin areas (Figure 8, Plate 12).

Questions

1. What type of symmetry is present in the annelids?

2. What is meant by a closed circulatory system? What is the adaptive advantage of such a system?

3. What substances are removed by the nephridia?

4. Explain the differences between seminal receptacles and seminal vesicles.

5. Without lungs or gills, how is respiration accomplished in the earthworm?

6. Of what economic significance are annelid members?

Studying the Animal Kingdom: Phylum Arthropoda

ANIMALS WITH JOINTED APPENDAGES: CHELICERATA AND CRUSTACEA

Learning Objectives

Explain how tagmatization in arthropods has been important in the creation of the arthropod body plan.

Explain how the exoskeleton has contributed to the success of the arthropods.

Explain how jointed appendages have contributed to the success of the arthropods.

Explain how the arthropod diet has been determined by the type of mouth appendages.

Identify the four main features of the chelicerates.

Identify the five main features of the crustacea.

INTRODUCTION

The phylum Arthropoda is a large and varied group of animals that constitute more than three-fourths of all known species of animals, approximately a million species. Success of the phylum is obvious, whether measured by total numbers, species, adaptability, or structural variety.

Although arthropod bodies are segmented like that of the annelids, they have achieved the most diversity in body form by dramatically changing and adapting to life on land. To reduce drying out, and provide protection and muscle anchorage, arthropods have a rigid **exoskeleton** made of chitin. Although chitin is found in other organisms, no other group uses this material as extensively as arthropods.

Second, the arthropod body has been regionalized into rigid, functional units or **tagmata.** This segregation forms distinctive **head, thorax** and **abdomen** regions. Third, from these body tagmata, specialized and **jointed appendages** cooperate in feeding, reproduction, and dramatic movement. Members of each subphylum and class modify the arthropod form into the diverse numbers present on the planet.

SUBPHYLUM CHELICERATA

Class Merostomata

Example: *horseshoe crab (Limulus)*

The more primitive of the arthropod phyla include some fascinating members such as the spiders, scorpions, ticks, mites and the horseshoe crab. All of these animals share four common features: (a) no antennae;

(b) no mandibles (chewing mouthparts); (c) anterior appendages are modified into **chelicera** and pedipalps; and (d) the body has two tagmata: a fused head/thorax, or **cephalothorax** and the abdomen.

Observe the shell of a horseshoe crab. Note the armored shell of the cephalothorax, the **prosoma,** and the covered abdomen (**opisthosoma**) including a spined tail or **telson.** Also observe the **compound eyes** on the dorsal surface of the prosoma. Turn the animal over and carefully look at the appendages. The first pair are pincered **chelicera** for manipulating food. The second are the longer **pedipalps** and the last four pair are **walking legs.** Horseshoe crabs belong to an ancient group of animal survivors. Early fossils, 500 million years old, are evidence that the body and lifestyle of the merostomata have been very successful (Figure 1, Plate 1).

FIGURE 1 Horseshoe Crab (Limulus).

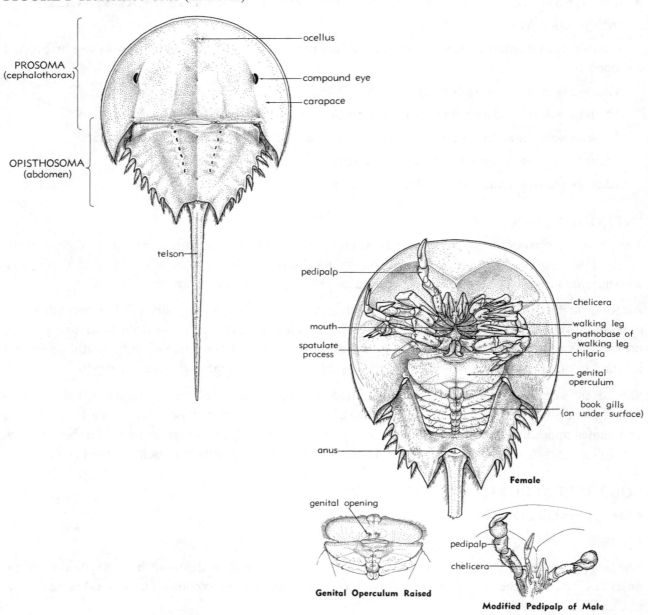

PLATE 1 Horseshoe Crab.

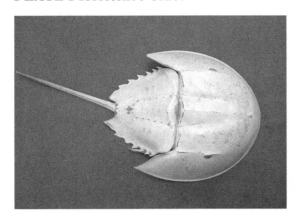

PLATE 2A Spider.

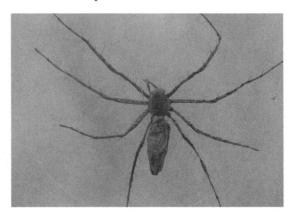

PLATE 2B Scorpion.

SUBPHYLUM CHELICERATA

Class Arachnida

The general body form of the chelicerates is modified in the arachnids and represented by the spiders, scorpions, ticks, mites. The fused **cephalothorax** and **abdomen** tagmata are retained in most but can be fused into one skeleton, as in mites and ticks. The **chelicera** vary from the small, manipulating appendages of scorpions and ticks to the terminal **poison** fangs in spiders. **Pedipalps** are usually food manipulating appendages but are dramatically modified in the scorpion's grasping claws. Four pairs of **walking legs** are also attached to the cephalothorax. The segmented abdomen may have the familiar poison **stinger** of scorpions or the **silk glands** and weaving **spinnerets** of spiders. Observe dried, jarred or table specimens of the arachnid group and identify specific examples and recognizable body parts (Figure 2, Plate 2A–B). If prepared slides are available of mites (*Sarcoptes, Demodex*) or ticks (*Dermacentor, Boophilus*), use the dissecting microscope and view their anatomy.

SUBPHYLUM CRUSTACEA: *CLASS MALACOSTRACA*

Example: Crayfish *(Astracus or Cambarus)*

Members of the subphylum Crustacea include the crayfish, crab, shrimp, water flea, barnacle, and other types. The great majority of crustaceans are aquatic or marine, while a few live on land. Crustaceans possess all of the characteristics of arthropods, but are distinguished from other classes of arthropods

FIGURE 2 Representative Chelicerates.

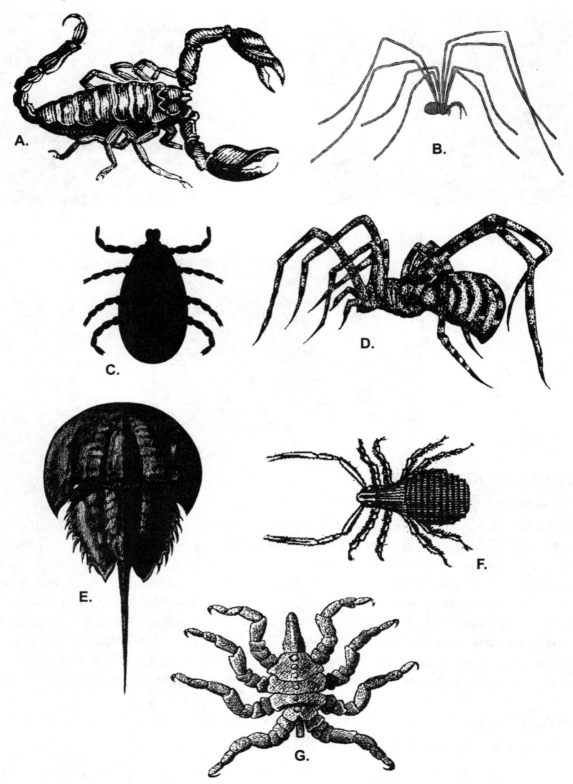

by having two pairs of **antennae,** two tagmata, the cephalothorax and abdomen, one pair of chewing **mandibles** and two pairs of oral **maxillae.** The appendages of crustacea are forked at the terminal end and are called **biramous.** Since most crustaceans are aquatic, respiration is by **gills** protected by the exoskeleton.

External Anatomy

Obtain a preserved crayfish, hand lens, and a dissecting pan. Place the crayfish in the pan in its usual upright position. Starting at the anterior end, identify the various parts. The **cephalothorax** is covered by a continuous **carapace** which acts as a protective shield. A **cervical groove** marks the division between the head and the thorax. At the anterior end of the carapace is a median pointed **rostrum.** On either side beneath the rostrum is a stalked, movable **compound eye.** Locate the two pairs of antennae: the smaller **antennules** and the longer **antennae.** The abdomen ends in a flared tail segment with one central **telson** and winged **uropods** (Figures 3, 4, 5, Plates 3A–B, 4A–B).

Turn the crayfish over so the ventral surface is now visible. Starting at the mouth fold back and locate the **mandibles,** two **maxillae** and three **maxillipeds** (respectively) around the mouth. The first walking leg pair is modified into the **cheliped.** Four other pairs of **walking legs** or **pereiopods** are used in locomotion and food manipulation. The abdominal appendages or **swimmerets** create water currents and help to carry the eggs on female bodies. Sexual differences can be seen in the first two pairs of swimmerets. In the male, the larger, stiffer, **gonopods** are modified for sperm transfer. In the female, the first two pairs of swimmerets are nondescript or greatly reduced and the opening of the **oviduct** is seen between walking legs #3 and #5. Finally, note the **anus** opening in the **telson** segment.

FIGURE 3 Crayfish (External).

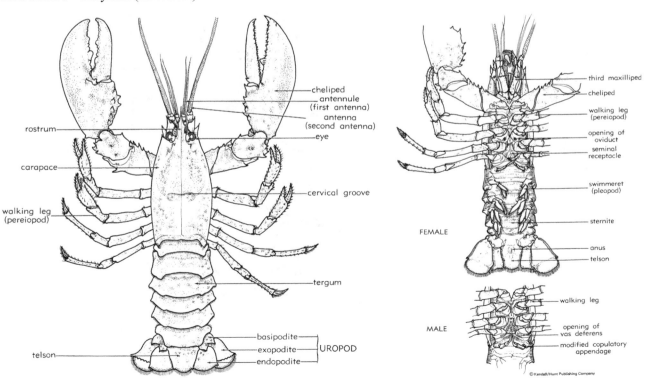

FIGURE 4 Crayfish (Lateral).

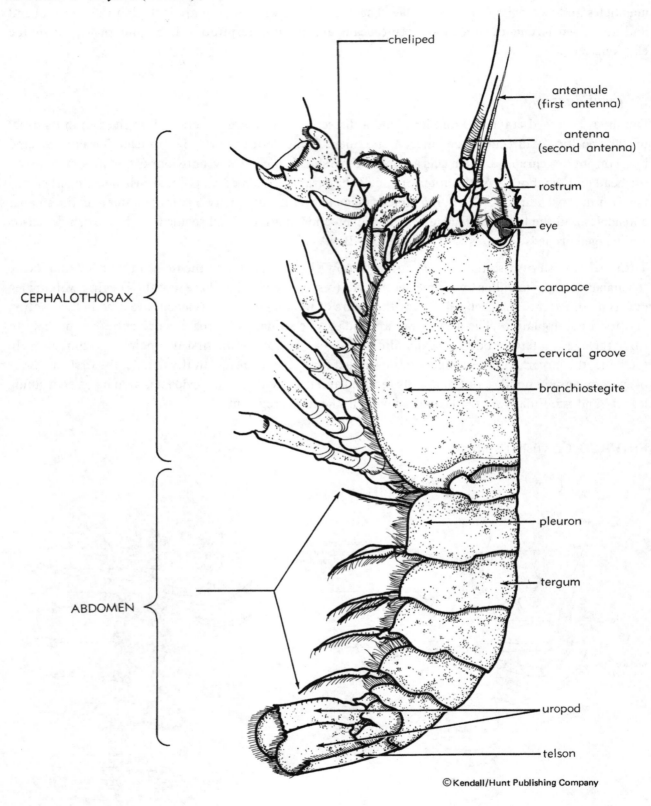

CEPHALOTHORAX

ABDOMEN

cheliped

antennule
(first antenna)

antenna
(second antenna)

rostrum

eye

carapace

cervical groove

branchiostegite

pleuron

tergum

uropod

telson

FIGURE 5 Crayfish (Varied Appendages).

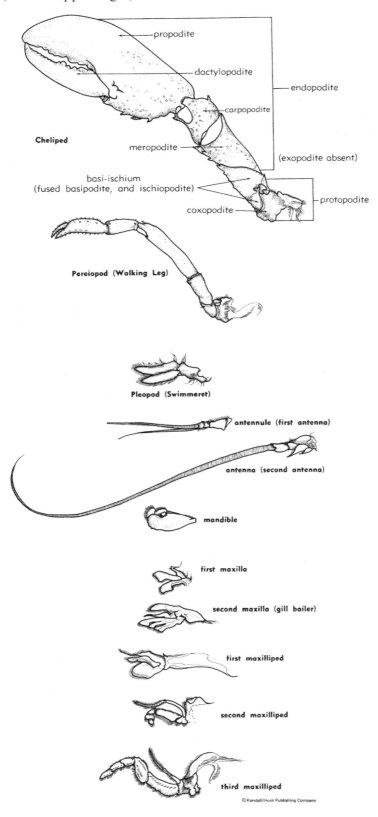

PLATE 3A Crayfish (Dorsal).

PLATE 3B Crayfish (Ventral).

PLATE 4A Crayfish (Female).

PLATE 4B Crayfish (Male).

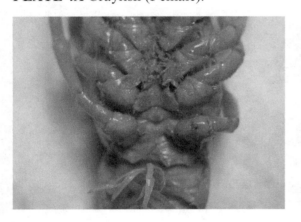

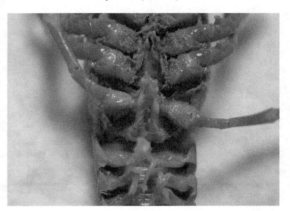

Internal Anatomy

Carefully remove the carapace by making an initial cut at the cephalothorax and abdomen joint and snipping along the mid-dorsal line **forward** to a point just behind the eyes. Leave the eyes in place. Gently free up the front edge of the carapace and separate the soft tissue away from the exoskeleton. Remove both sides of the carapace. The soft, feathery **gills** cover both sides. Directly behind the eyes, locate the double pouched **stomach (cardiac and pyloric)** with a curved **gastric mill** forming a rigid ring. On either side should be a light mass of tissue, the **digestive glands** that aid in the digestive process. The shield-like **heart** is centrally located behind the stomach. Posterior and lateral to the heart are the paired **gonads.**

Gently remove the stomach, heart and digestive glands and locate two circular, plate-like **green glands** on either side of the mouth opening. These antennal or maxillary glands are used for osmoregulation. Two white, thin **nerve tracts** point anteriorly to the larger ganglion (**brain**) (Figures 6, 7, Plate 5).

Pointing the scissors posteriorly, now continue a center cut through the abdominal shell and peel it away. Identify the central tube, **intestine,** and the large flexor and extensor **muscles.**

Other Crustacea

Examine other types of crustaceans on the demonstration table (crab, shrimp, barnacles, etc.). Turn over each sample and identify the characteristics that places them in the crustacean group.

FIGURE 6 Crayfish (Dissection).

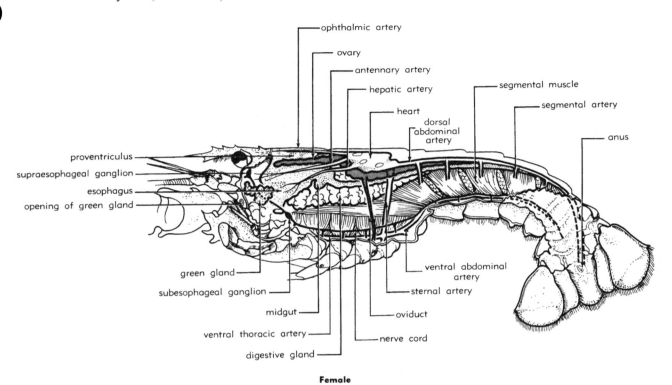

Female

LEFT DIGESTIVE GLAND REMOVED; TAIL TURNED TO SHOW VENTRAL SURFACE

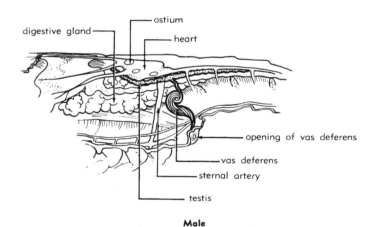

Male

AREA OF GONADS; LEFT DIGESTIVE GLAND REMOVED

FIGURE 7 Crayfish Dissection (Dorsal View).

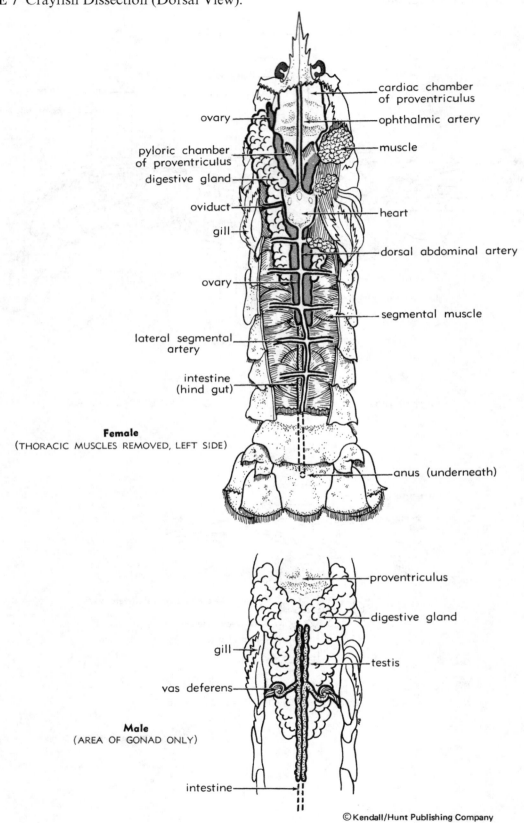

cardiac chamber
of proventriculus

ovary

ophthalmic artery

pyloric chamber
of proventriculus

muscle

digestive gland

oviduct

heart

gill

dorsal abdominal artery

ovary

segmental muscle

lateral segmental
artery

intestine
(hind gut)

Female
(THORACIC MUSCLES REMOVED, LEFT SIDE)

anus (underneath)

proventriculus

digestive gland

gill

testis

vas deferens

Male
(AREA OF GONAD ONLY)

intestine

PLATE 5 Crayfish (Internal).

Questions

1. Name some crustaceans that might be found in a terrestrial habitat.

2. What sensory organs does the crayfish have that inform it about its surroundings? Which of these would be helpful in muddy water?

3. What is the function of the different types of crayfish appendages?

4. Of what medical and economic significance are the arachnids?

5. Of what economic significance are the crustaceans?

ANIMALS WITH JOINTED APPENDAGES: SUBPHYLUM UNIRAMIA
Learning Objectives

Identify the physical characteristics of insects that separate them from the other arthropods.

Describe how insects have solved the problem of survival on dry land.

Recognize members of various insect orders.

Identify the positive and negative impact that insects have on the planet.

INTRODUCTION

The Uniramia are identified as arthropods whose appendages have only one (uni) branch. They possess the basic characteristics of exoskeleton, jointed appendages and tagmata that have been seen in the chelicerata and crustacea. The insects have exploded in numbers and varieties by adapting well to terrestrial life. They have improved on the basic plan with well developed sensory organs, unparalleled reproductive fecundity and varied their mouthparts to exploit all types of food. Furthermore, they have complicated behavior patterns, developed social orders, undergo metamorphosis and most dramatically achieved flight.

CLASS CHILOPODA

The centipedes are land arthropods with flattened bodies. The head appendages are similar to the insects and their somite number can range from a few to over 100. Each segment, excluding the first and last two, bear a pair of jointed legs. The first pair of these "hundred-leggers" are modified into poison claws. The centipedes search for prey in moist places, such as under logs and stones, paralyze them with the poison claws and chew them with mandibles. Observe preserved centipedes from the display table. Using a dissecting scope, focus on the anterior end and look for the modified head appendages (Plate 6).

CLASS DIPLOPODA

Millipedes with 11 to 100 segments differ from the centipedes by having two pairs of appendages on each somite. More rounded in body form, their "thousand-legs" propel them quickly yet strongly through ground litter to feed on plant matter. Many of these herbivores also possess **repugnatorial glands** capable of spraying hydrogen cyanide when threatened. Compare preserved millipedes to centipedes and be prepared to tell them apart (Plate 7).

PLATE 6 Centipede.

PLATE 7 Millipede.

CLASS INSECTA

Example: *Romalea microptera*

External Anatomy

Obtain a preserved grasshopper, hand lens and a dissecting pan. Place the grasshopper in the pan in its usual upright position. Starting at the anterior end, identify the various parts. Note the body is divided into three tagmata: **head, thorax** and **abdomen.** On the head tagmata, locate the single pair of **antennae,** one pair of **compound** eyes, three simple eyes or **ocelli** and the mouth parts. Remove each mouth part by grasping the piece at its base with forceps and pulling sharply. These include the upper **labrum** with its base clypeus, two hard, serrated **mandibles,** two **maxillae** with palps and a lower **labium.** The hypopharynx remains as the last piece in the center of the mouth (Figures 8, 9, 10).

The thorax is composed of three fused somites. On the first, **prothorax,** is attached a pair of **walking legs.** A second pair of walking legs and a pair of **forewings** or protective wings are attached to the mesothorax. Finally a pair of jumping legs and a pair of flying or **hindwings** are attached to the **metathorax.**

Note the abdominal segments. The first one has an oval window or **tympanum** that functions in hearing. The last three segments are modified for reproductive purposes. In the female spiny **ovipositors** are utilized for egg deposit; in the male tail segments are modified for copulation. Along either side of the abdomen, locate very small openings or **spiracles** which are external openings for respiration.

Identify the individual leg parts. Starting from the attachment point, the **coxa** and **trochanter** form a rotating cuff and articulates with the **femur.** The **tibia** is spiny and terminates in the flexible and clawed **tarsus.** Although the legs have similar parts, the hindleg is larger and obviously modified for jumping.

Internal Anatomy

Remove the legs and wings of the grasshopper. Make a shallow incision in the tip of the abdomen and cut along the mid-dorsal line to just behind the head. Peel away the exoskeleton, being careful to tease away the body tissues. Just anterior to the head, locate the **crop,** used for food storage. Finger-like projections or the **gastric caeca** of the stomach increase the surface area for food absorption. Along the dorsal surface, under a thin membrane, lies the slightly enlarged **heart** and slender, tubular, dorsal **aorta.** **Malpighian tubules** or excretory organs are thin, filamentous projections closer to the tail. In females, oblong **eggs in ovaries** are tightly packed along the dorsal surface of the digestive system. The **oviducts** connect the ovaries to the **vagina** opening. In males, a pair of **testes** are bound together above the digestive tract. Beneath them are the two **sperm ducts** (vas deferens) that meet to form a single **ejaculatory duct** (Figure 11A, B, C).

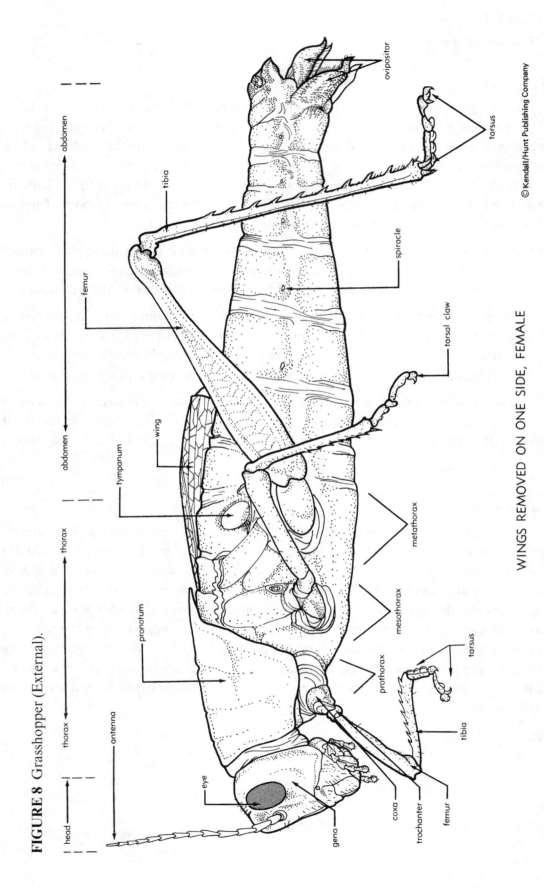

FIGURE 8 Grasshopper (External).

WINGS REMOVED ON ONE SIDE, FEMALE

FIGURE 9 Grasshopper (External detail).

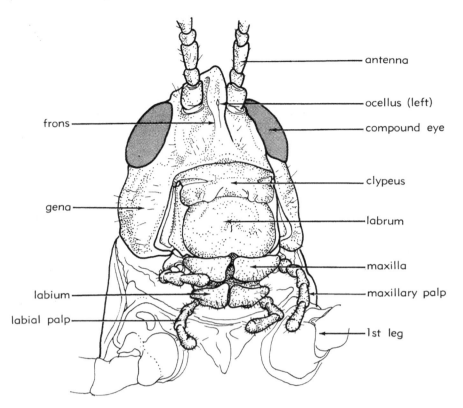

Head, Ventral View

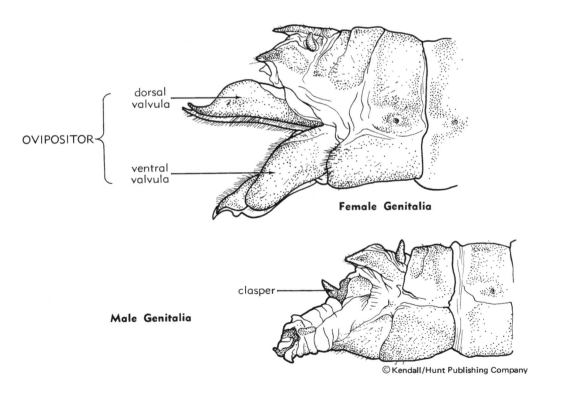

Female Genitalia

Male Genitalia

FIGURE 10 Grasshopper mouth parts.

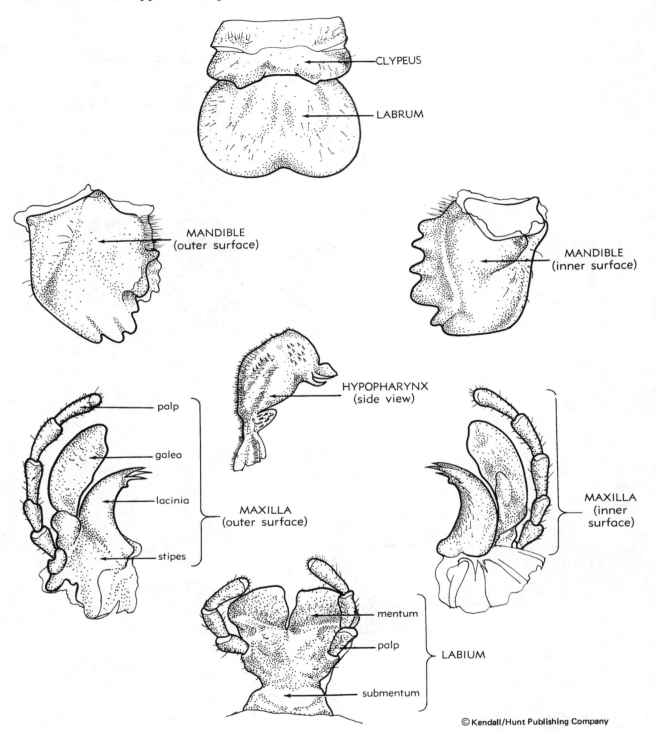

CLYPEUS

LABRUM

MANDIBLE (outer surface)

MANDIBLE (inner surface)

HYPOPHARYNX (side view)

palp

galea

lacinia

stipes

MAXILLA (outer surface)

MAXILLA (inner surface)

mentum

palp

submentum

LABIUM

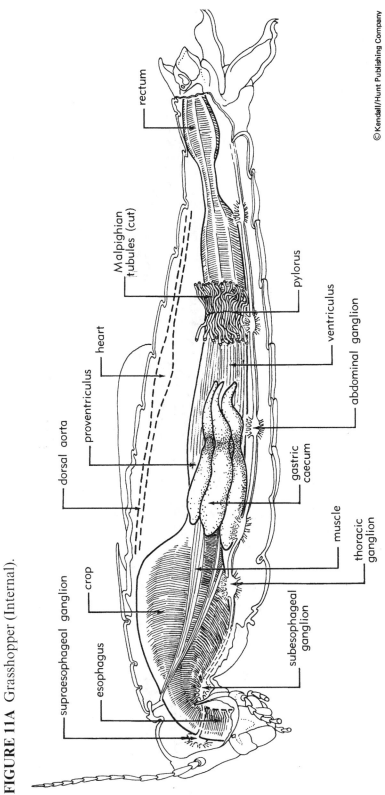

FIGURE 11A Grasshopper (Internal).

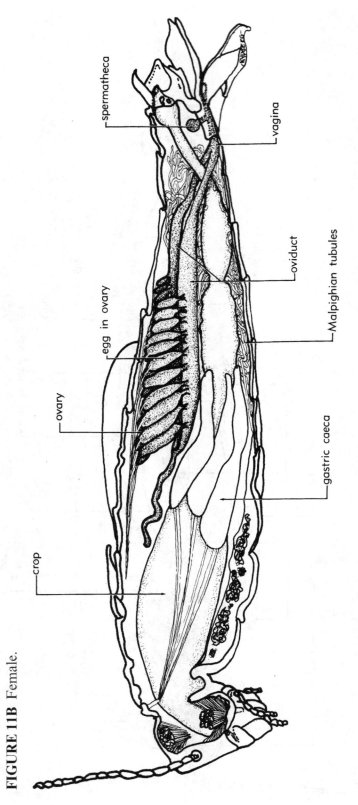

FIGURE 11B Female.

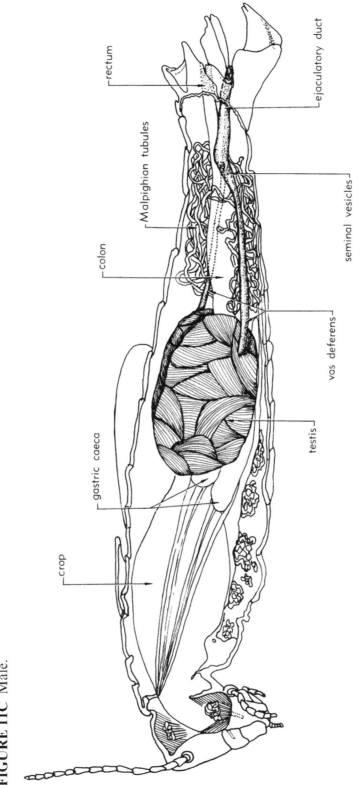

FIGURE 11C Male.

Questions

1. What is the difference between compound and simple eyes?

2. How do the two pairs of wings differ in the insect orders?

3. Give at least 4 characteristics of insects that have made them so successful in a terrestrial environment.

4. In what ways do millipedes and centipedes differ from each other?

5. What positive impact do the insects have on the planet?

6. What negative impact do the insects have on the planet?

WORKSHEET TO DISTINGUISH UNIRAMIA AND INSECT ORDERS

The following worksheet identifies and describes selected examples of the Subphylum Uniramia with focus on the insect orders. Collect examples of various insects or use preserved specimens to determine their names and orders. Use the dichotomous key to complete this project.

PHYLUM ARTHROPODA

The Uniramians

Subphylum Uniramia

Class Chilopoda *(centipedes)*

Examine those centipedes on display. Note the flattened trunk with *a pair of appendages for each segment.* Also observe the large **fangs** located on the maxillipeds just ventral to the elongated antennae. Finally, examine the sensory **cerci** projecting from the final segment.

Class Diplopoda *(millipedes)*

Compare the basic structure of a millipede with that of a centipede. Note the rounded shape, lack of poison fangs, and presence of *two pairs of appendages* for each body diplosegment.

Class Insecta *(insects)*

There are a number of insects on display. Examine each to learn its specific properties. Note the three body regions **(head, thorax,** and **abdomen)**, *three pairs of legs* borne on the thorax, and wings (if present). Use the key provided to identify each of the orders portrayed below. Using a dissecting microscope, or a magnifying glass, compare each to the description given.

REPRESENTATIVE INSECT ORDERS

Ephemeroptera *(mayflies)*

Mayflies are primitive insects with a prominent aquatic larval stage. The adult part of life cycle is reduced; adults live for only a few days, at most, during which time they don't feed. They mate in swarms while flying. When at rest, the wings are held vertically over the back; the abdomen has three long thread-like cerci, legs are usually small, mouthparts are pronounced in larva but vestigial in adults. Mayflies are an important food source for many fish.

Odonata *(dragonflies and damselflies)*

Aquatic larvae are predacious, as are the adults, which prey on flying insects. These primitive insects have large compound eyes, reduced, bristlelike antennae, long and slender abdomens, chewing mouthparts, and two pairs of net-veined wings held out to the side when at rest (dragonflies) or vertically over the back (damselflies).

Orthoptera *(grasshoppers, cockroaches, katydids, crickets, walking sticks, mantids)*

Legs adapted for running (cockroaches), jumping (grasshoppers, crickets, katydids), or raptorial (mantids). Forewings narrow and leathery, hind wings usually large and fan-like, chewing mouthparts, cerci present, antennae moderate to very long, female with pronounced terminal ovipositor, sound, reception is a tympanum on legs or abdomen, and parts of the exoskeleton may stridulate to produce sound.

Isoptera *(termites)*

All termites are eusocial with a well-defined caste system of blind workers, blind soldiers with enlarged mandibles, and reproductives (males are winged but live for only a brief period, queens are initially winged, but shed them to become large reproductive units cared for by workers). These are small, soft-bodied insects. The abdomen is broadly joined to thorax, they live in huge swarming colonies, highly destructive when they occur in human residences.

termite queen

Dermaptera *(earwigs)*

In some locations, earwigs are common household residents, but they are scavengers and present no health threat. They possess enlarged forceplike abdominal cerci used in defense, hind wings are large and folded under short, horny forewings.

Anoplura *(sucking lice)*

These are blood-sucking ectoparasites of birds and mammals, including humans. Head lice (cooties) and pubic lice (crabs) have infested mankind throughout history. They have piercing-sucking mouthparts, wings are absent, legs adapted for attachment, compound eyes missing, broad body capable of great distention during a blood meal.

Hemiptera *(true bugs)*

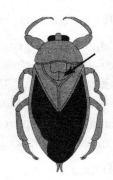

Piercing-sucking mouthparts form a jointed needlelike beak, that arises from the front portion of the head and is held between the forelegs when not in use. Both sets of wings lie flat against the abdomen when at rest; forewings thick and leathery at the base and membranous toward the tip, hind wings large and membranous but folded beneath the forewings. Prominent triangular plate (the scutellum) located between the bases of the forewings. Most are herbivores, many are predacious, some are ectoparasites (kissing bugs and bed bugs both infest humans). As a group, hemipterans are of great economic importance as destroyers of crops.

Homoptera *(cicadas, leafhoppers, plant hoppers, aphids)*

Feed on plant sap, mouthparts modified for piercing and sucking. Some forms are wingless but most have two pair of membranous wings held tentlike or rooflike over the abdomen when at rest.

Hymenoptera *(ants, bees, and wasps)*

This order is characterized by the presence of a narrow waist that joins the abdomen to the thorax. Chewing mouthparts, some have a stinger derived from the ovipositor, hind wings small and joined to larger forewings by a hook. Tendency in this group to form eusocial units. Some species are nuisances, others are highly beneficial in pollinating plants.

Siphonoptera *(fleas)*

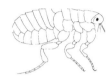

Small, wingless insects, body laterally compressed, mouthparts modified to feed exclusively on blood, hindlegs adapted for jumping, other legs for clinging. All are ectoparasites and some are transmitters of serious disease (i.e., bubonic plague). Adults are not highly specific and will switch from one host species to another; humans are often infested with dog, cat, or rat fleas, as well as their own.

Diptera *(flies, mosquitoes, and gnats)*

One pair of wings is usually present, the other pair reduced to knobby structures called halteres used as gyroscopes for balance while flying, large eyes, highly mobile head. Mouthparts are variable and may be adapted for sponging, piercing, or chewing. Of great economic and medical importance; nuisance Lepidoptera (butterflies and moths). Large compound eyes with long antennae, wings and body covered with overlapping, dense, pigmented scales. Mouthparts modified into a long, coiled tube to extract nectar. Butterflies hold wings vertically at rest, moth, horizontally. Larvae of great economic importance as destroyers of crops and cloth; adults are important pollinators.

Coleoptera *(beetles)*

This is the largest order of insects; there are more beetles than all non-arthropods combined. Head bears well-developed antennae and eyes, two pairs of wings with the first pair consisting of a hard shell-like covering, the elytra, that meets in a straight line down the back, hind wings membranous and folded under forewings when at rest, body compact and hard, mouthparts chewing. Some are serious agricultural pests, while others are predators of harmful pests.

KEY OF SELECTED INSECT ORDERS*

1. a. Wings present and well developed 2
   b. Wings absent 20

2. a. Forewings thickened or leathery (at least at base); hind wings membranous, may be hidden beneath forewings 3
   b. All wings membranous throughout 7

3. a. Mouthparts beaklike (somewhat like a syringe and often held between legs when not in use) 4
   b. Mouthparts mandibulate, adapted for chewing 5

4. a. Beak arises from front of head, forewings leathery at base but membranous at tips (true bugs) . . . . . **Hemiptera**
   b. Beak arising from rear of head, forewings uniform and held tentlike over abdomen (plant hoppers) . . . . . **Homoptera**

5. a. Abdomen with forcepslike cerci, forewings short and covering folded hind wings (earwigs) . . . . . **Dermaptera**
   b. Not as above 6

6. a. Forewings hard, veinless, shell-like meeting in a straight line; hind wings folded under forewings (beetles) . . . . . **Coleoptera**
   b. Forewings not as above, veined, hind wings broad and usually shorter than forewings; forelegs may be raptorial, hind legs may be modified for jumping (grasshoppers, mantids, crickets, cockroaches) . . . . . **Orthoptera**

7. a. One pair of wings only 8
   b. Two pair of wings 11

8. a. Body grasshopper-like, hind legs adapted for jumping . . . . . **Orthoptera**
   b. Not as above 9

9. a. Mouthparts vestigial (missing); three long, threadlike cerci (earwigs) . . . . . **Dermaptera**
   b. Mouthparts chewing, no cerci 10

10. a. Sucking mouthparts; hind wings reduced to clublike halteres, tarsi five segmented (flies) . . . . . **Diptera**
    b. Mouthparts chewing, tarsi two or three segmented . . . . . **Psocoptera**

11. a. Wings covered by overlapping, pigmented scales; mouthparts coiled (butterflies/moths) . . . . . **Lepidoptera**
    b. Not as above 12

12. a. Wings long and narrow, fringed with extremely long hairs (edges appear fuzzy) . . . . . **Thysanoptera**
    b. Not as above 13

13. a. Forewings large and triangular; hind wings small and rounded; wings held vertically over back; wings heavily veined; soft-bodied insects with two or three long, threadlike tails (mayflies) . . . . . **Ephemeroptera**
    b. Not as above 14

14. a. Tarsi five segmented (just count the segments on the distal leg region [tarsus]) 15
    b. Tarsi consisting of four or fewer segments 16

---

*Modified from Borrer and DeLong, An Introduction to the Study of Insects. Holt, Rinehart & Winston.

15. a. Rather hard-bodied, wasplike insects; usually a narrow waist attaches abdomen to thorax; hind wing smaller than forewing . . . . . (bees, wasps, yellowjackets) . . . . . **Hymenoptera**

    b. Not as above 16

16. a. Hind wings and forewings of equal size; wings with many veins; antennae short and bristlelike; abdomen long and slender; wings outstretched at rest (dragonflies) or held vertically (damselflies) . . . . . **Odonata**

    b. Not as above 17

17. a. Sucking mouthparts (mouthparts basically a tube or stylet) 18

    b. Chewing mouthparts 19

18. a. Beak arising from front of head (true bugs) . . . . . **Hemiptera**

    b. Beak arising from hind part of head (cicadas, aphids) . . . . . **Homoptera**

19. a. Wings similar to each other; soft bodied; cerci small, winged termites . . . . . **Isoptera**

    b. Not as above 20

20. a. Ectoparasites of birds and mammals; body flattened laterally or dorso-ventrally 21

    b. Not as above; soft bodied, usually blind insects, may have large pincerlike mandibles (wingless termites) . . . . . **Isoptera**

21. a. Body flattened laterally; jumping hind legs; body sparsely covered with spine or hairs (fleas) . . . . . **Siphonaptera**

    b. Body flattened dorso-ventrally; large tarsal claws for attachment to host (sucking lice) . . . . . **Anoplura**

Studying the Animal Kingdom: Phyla Echinodermata, Hemichordata, and Chordata

SPINY-SKIN ANIMALS: *WATER VASCULAR SYSTEM*

Learning Objectives

Describe the general structure of the echinoderm body plan.

Describe the structure and function of the water vascular system.

Recognize representative specimens of the various classes of echinoderms.

INTRODUCTION

The phylum Echinodermata includes sea stars, sea urchins, brittle stars, sea cucumbers, and feather stars. They form a large group of specialized marine organisms that are interesting because of their bizarre body form and their evolutionary relationship to the chordates. The echinoderm larvae undergo a striking change during their embryonic development from a form with bilateral symmetry to one with **radial symmetry.** The resulting symmetry, *pentaradiate* symmetry, always has the body parts arranged radially in fives or multiples of five. Apparently this type of symmetry suits the aquatic environment but reduces the need for major organ systems. Thus, they lack gills or lungs, a brain, a heart, kidneys, a distinctive head region and rely on external fertilization.

A unique characteristic of echinoderms is the **water vascular** system, that consists of numerous water-filled tubes ending in a large number of **tube feet.** They have: a coelomic cavity lined with ciliated peritoneum, an **endoskeleton** of calcareous plates, a spiny epidermis, dermal papillae as respiratory organs and a simple nervous system.

The echinoderms are **deuterostomes** like the higher organisms; namely the chaetognatha, hemichordates and the chordates. The embryology of sea stars was observed in Exercise 6. Review the development of the gastrula and determine why the echinoderms are placed in the deuterostome line, linking them in a common origin to the higher organisms.

CLASS ASTEROIDEA

Example: *Asterias*

External Anatomy

This class includes sea stars, or more commonly called starfish. The lab specimens have five rays or arms; however, some species may have six, seven, eleven, or up to 50 in number. Obtain a preserved sea star and place in a dissecting pan. Make a distinction between the **oral** (ventral) and **aboral** (dorsal) surfaces (Figures 1, 2, 3, Plate 1). On the aboral surface locate the button-like structure, **madreporite,** on the central portion called the **disc.** The disc has five **arms** or *rays* attached. The madreporite is between two arms, called the **bivium,** while the remaining arms are called the **trivium.**

With the aid of a hand lens or dissecting microscope, find little projections which form a ring around the base of each spine. These are called **pedicellariae,** pincher-like structures which keep the surface area free from debris. Also, distinguish soft, finger-like projections between the spines, these are the **dermal papillae.**

Internal Anatomy

To dissect the specimen, cut off the end of a **trivium** about one-half inch from the tip with a pair of scissors. Then make **two lateral** cuts extending toward the disc. Remove the body wall from the dorsal surface of the disc by cutting around its margins just inside the arms. Cut around the madreporite, **leaving it in place on the disc.** Gently lift up the dorsal wall of the arm and disc and gently scrape the attached organs into the bottom portion of each arm. The greenish structures that adhere to the arm coverings are the **pyloric caeca** (also called **hepatic caeca** or **digestive glands**) (Figure 2A). The sac-like **stomach** (cardiac and pyloric portions) sits in the center of disc and is distendable out the mouth area. A short **intestine** leads to the **anal** opening on the aboral surface, both of which are difficult to see.

A pair of pinkish, glandular **gonads** lie under the pyloric caecae in each arm. Gonads vary in size from very small to those filling the arm's cavity. The sexes are separate; however, to determine the sex, a microscopic examination of the gonadal tissue must be made.

Remove the pyloric caecae and gonads. Note the bulb shaped **ampullae** between the **limy ossicles** (Figure 2). Each tube foot is located in the **ambulacral groove.** The ampullae in each arm connect by **lateral canals** to a centrally located **radial canal.** Move to the central disc and remove the pyloric and cardiac stomachs. Be careful not to damage the **stone canal** which leads from the madreporite to the **ring canal** which circles the mouth region.

CLASS OPHIUROIDEA

Elongated, spindly arms about a small central disc identify the brittle stars of this class. The extended appendages gives them a graceful, yet wiry locomotion. The tendency of the arms to fragment earns them their common name, while serpent stars identifies their wriggly motion (Figure 3, Plate 2).

CLASS ECHINOIDEA

Sea urchins and sand dollars lack rays, but have many movable spines about a hemispherical body. The gonads of some sea urchins are considered to be delicacies in some island cultures. On the oral surface, five pointed teeth form the **Aristotle's lantern** around the mouth. The flattened forms of sand dollars and sea biscuits are covered by downy numerous spines. Once cleaned, their skeletons are collectable items in many beach stores (Figure 3, Plate 3, Plate 4).

FIGURE 1 Starfish (External).

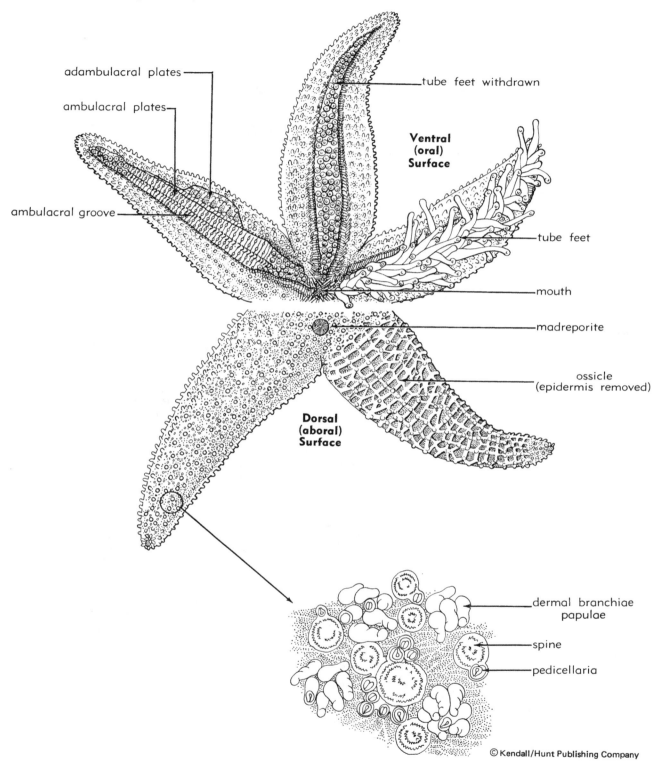

adambulacral plates

ambulacral plates

ambulacral groove

tube feet withdrawn

**Ventral
(oral)
Surface**

tube feet

mouth

madreporite

ossicle
(epidermis removed)

**Dorsal
(aboral)
Surface**

dermal branchiae
papulae

spine

pedicellaria

© Kendall/Hunt Publishing Company

Magnified Area on Dorsal Surface

FIGURE 2 Starfish (Internal).

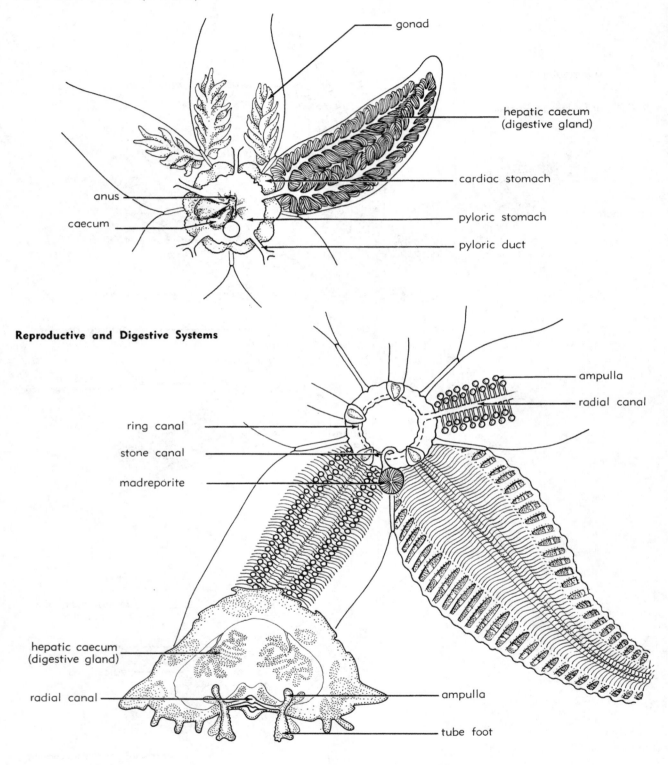

gonad

hepatic caecum
(digestive gland)

cardiac stomach

pyloric stomach

pyloric duct

anus

caecum

Reproductive and Digestive Systems

ampulla

radial canal

ring canal

stone canal

madreporite

hepatic caecum
(digestive gland)

radial canal

ampulla

tube foot

Water Vascular System

FIGURE 3 Representative Echinoderm Classes.

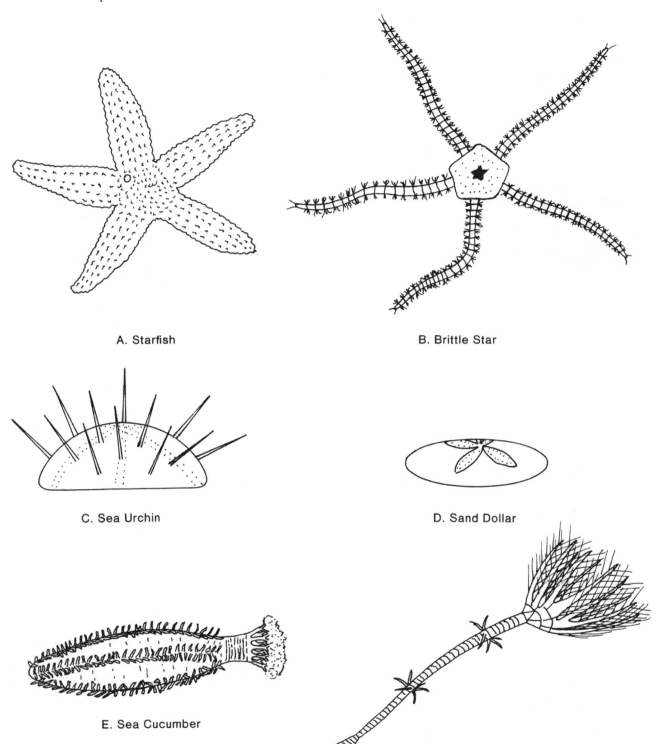

A. Starfish

B. Brittle Star

C. Sea Urchin

D. Sand Dollar

E. Sea Cucumber

F. Sea Lily

CLASS HOLOTHUROIDEA

The class Holothuroidea is represented by the sausage-shaped sea cucumbers. They have a slender elongated body on an oral-aboral axis that is protected by a leathery epidermis. Tentacles around the mouth filter the water for food. Oddly, the sea cucumbers are capable of **evisceration** of body parts when threatened. Hiding away, they will regenerate the soft inner parts left behind (Figure 4, Plate 5).

PLATE 1A Starfish (aboral surface).

PLATE 1B Starfish (oral surface).

PLATE 1C Starfish Anatomy.

PLATE 2 Brittle Star.

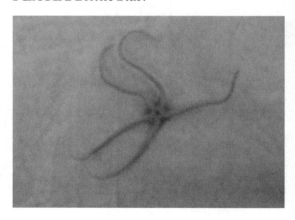

PLATE 3 Sea Urchin.

PLATE 4 Sand Dollar.

FIGURE 4 Sea Cucumber (External).

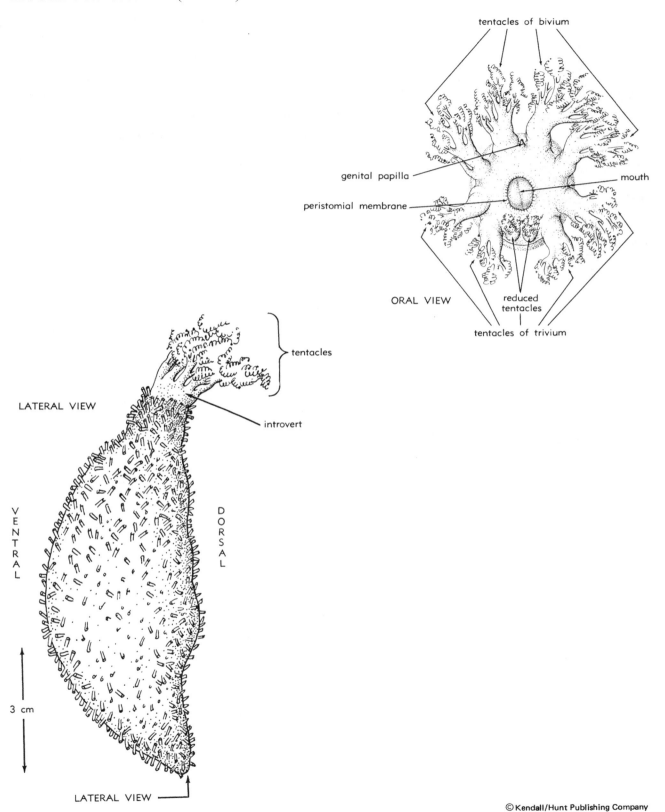

PLATE 5 Sea Cucumber.

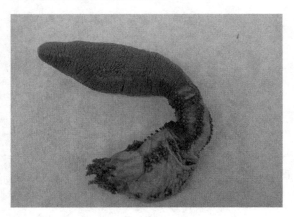

CLASS CRINOIDEA

The class Crinoidea includes the sea lilies and feather stars. These echinoderms remain attached for much of their lives. Sea lilies have a flower-shaped body at the tip of an attached stalk while the feather stars have long multi-branched arms with no stalk. Crinoid fossils date back to the Cambrian Period, evidence of the lasting power of these fragile looking echinoderms.

Examples of each of the echinoderm classes are available on the display table. Identify the primary features that place them in each respective class.

Questions

1. Do sea stars have an anterior end?

2. Of what use are the pedicellariae?

3. Are the spines a portion of an endoskeleton or exoskeleton?

4. Explain the function of the pyloric caeca.

5. How are the tube feet able to pull open bivalve molluscs?

6. Explain the physiology of the water-vascular system.

7. Of what economic impact are the echinoderms?

PHYLUM CHORDATA

Learning Objectives

Describe the 4 basic characteristics of the chordate phylum.

Identify the primary characteristics which separate invertebrates, chordates and vertebrates.

Recognize members of each of the vertebrate classes.

INTRODUCTION

The invertebrate animals, previously studied, represent successful members of various lower phyla. Whether simple or complex; soft-bodied or protected by a skeletal structure; sessile or locomotive, they lack the major features of the most highly evolved phylum, the **Chordata.** At some time in their life cycle, members of this phylum possess the following: (1) a stiffening rod-like structure or **notochord,** (2) a **dorsal, hollow nerve cord,** (3) paired, pharyngeal **gill slits,** and (4) a **postanal tail.**

Chordates also retain the familiar features of: bilateral symmetry, complete digestive tract, segmentation, cephalization and deuterostome development. Three subphyla (Urochordata, Cephalochordata and Vertebrata) contain an array of organisms whose embryological development is very similar but whose adult forms vary dramatically. Their organ-systems have evolved as life emerged from aquatic to terrestrial environments. Sometimes referred to as the **protochordates** (proto; "first"), both urochordates and cephalochordates are the nonvertebrate groups. The Craniata or Vertebrata possess one more structural feature: a bony or cartilaginous **skeleton** surrounding the nervous system. Hence, the name derives from the protection around the brain (**cranium**) and the spinal cord (**vertebrae**).

SUBPHYLUM UROCHORDATA

The urochordates are unique animals that live in the sea either as solitary organisms or in colonial forms. Metamorphosis transforms a free-swimming, tadpole-like body with the four chordate characteristics into

PLATE 6 Urochordata (Sea Squirt).

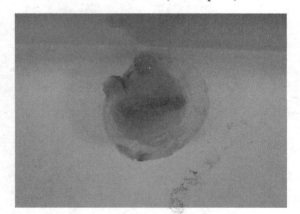

a sac-shaped bag usually attached to the ocean floor. Observe a specimen of an adult sea squirt (*Molgula*). Note the delicate appearance of its covering, the **tunic,** composed of a polysaccharide similar to cellulose. Locate two, tubular **siphons:** the incurrent brings a water stream into a modified pharynx having gill slits while the excurrent forcefully discharges the filtered water. Because of these features, urochordates are also known as **tunicates** or **sea squirts.** The nervous system is reduced to a single elongated neural ganglion near the incurrent siphon. There is no evidence of a notochord or post-anal tail in the adult form (Plate 6).

Larval sea squirts are free-swimming with the stiffening skeletal rod or **notochord** confined to the tail extension. A dorsal nerve cord runs along the length of the notochord. Although there is no distinctive head region, food and water enter the mouth and pass into the slitted pharynx. Water flow exits a second opening, the **atriopore.** The dramatic transformation from larva to adult results in an equally dramatic loss of most of the chordate characteristics. Lacking a skeletal covering around the nerve cord identifies the urochordates as invertebrates.

SUBPHYLUM CEPHALOCHORDATA

Example: *Branchiostoma (Amphioxus)*

The **lancelet,** *Branchiostoma* or *Amphioxus,* represents the cephalochordates; so-called because the notochord extends from the head to the tail region. They represent an evolutionary stage in which chordate characteristics are retained throughout the entire life, but vertebrate characteristics are absent. Living along sandy shores, their transparent bodies filter detritus through prominent gill slits. Obtain a preserved lancelet and scan its length using a hand lens or dissecting scope. Note the anterior end has a fringed covering of **cirri** (sing. cirrus) which contains chemoreceptors and helps filter large particles. Water enters the mouth, which is invaginated into a protective chamber or **oral hood.** It exits through the **atriopore,** an opening about two-thirds down the length of the animal body. Distinctive V-shaped blocks of muscle, the **myotomes,** are located on both sides of the body. Below the myotomes, mature **gonads** are visible through the epidermis. Extended dorsal and ventral **fins** can be traced posteriorly to the **caudal fin** forming the **post-anal tail** (Figure 5, Plate 7).

In a cross section slide of *Branchiostoma,* locate the dorsal fin ray and progress ventrally to identify the: nerve cord, notochord and chambered pharynx. The lateral pharyngeal walls have supporting **gill bars** separating the open **gill slits.** In a cut made at the pharyngeal level, the **myotome** segments progress in block-like form. Surrounding the pharynx but on the ventral surface, locate the two large **gonad** chambers (Figure 6).

SUBPHYLUM VERTEBRATA

The addition of a cartilaginous or bony skeleton around the nervous system clearly separates and identifies the vertebrate animals. Ranging from the elongated bodies of the agnaths to the four-limbed mammals, vertebrates are found in all types of habitats. They have modified their organ-systems for life in the water, on land and even for flight.

SUPERCLASS AGNATHA

Jawless Fish

Hagfish and **lampreys** have eel-like bodies with a cartilaginous skeleton and a persistent notochord. They lack jaws but their mouths may be surrounded by well developed teeth or suckers. Locate specimens of both on the display table. Marine hagfish burrow in the mud feeding primarily on polychaete worms and dead fish. Their integumentary system has large secretory glands capable of producing copious amounts of slimy **mucus.** Lampreys use **suckers** and cutting **teeth** to attach to fish for a blood meal. Marine lampreys are studied for their migratory routes from salt water to fresh water streams for spawning. Look at the mouth area of preserved agnaths and note their primitive fish form (Plate 8).

PLATE 7 Cephalochordate (Amphioxus).

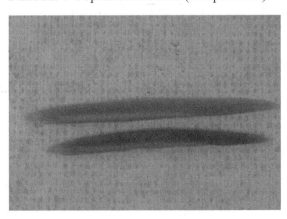

SUPERCLASS GNATHOSTOMATA (JAWED VERTEBRATES)
Class Chondrichthyes

The sharks, skates and rays are included in the chondrichthyes because of their **cartilaginous** skeleton. Paired **fins, heterocercal** tail and a **ventral mouth** help to distinguish these fish from the bony ones. Another external

FIGURE 5 Amphioxus (External View).

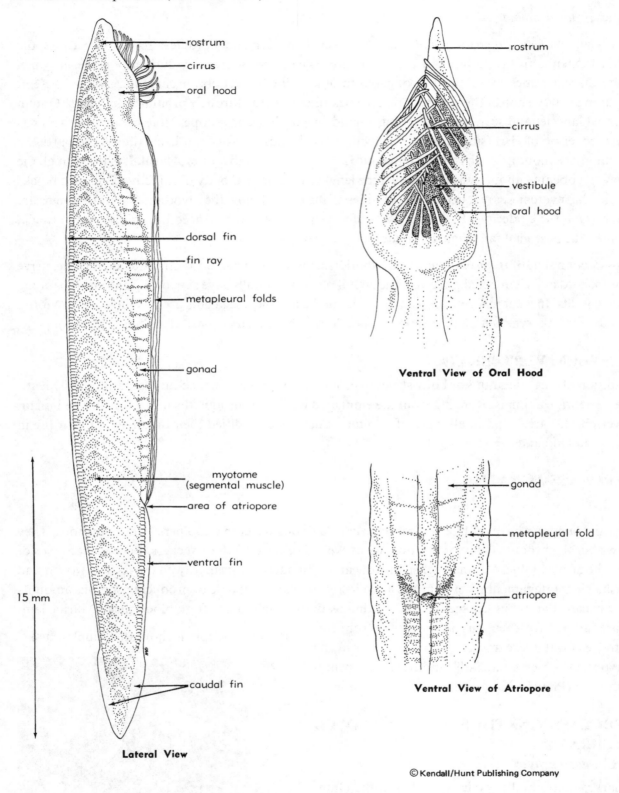

Ventral View of Oral Hood

Ventral View of Atriopore

Lateral View

FIGURE 6 Amphioxus (Internal View).

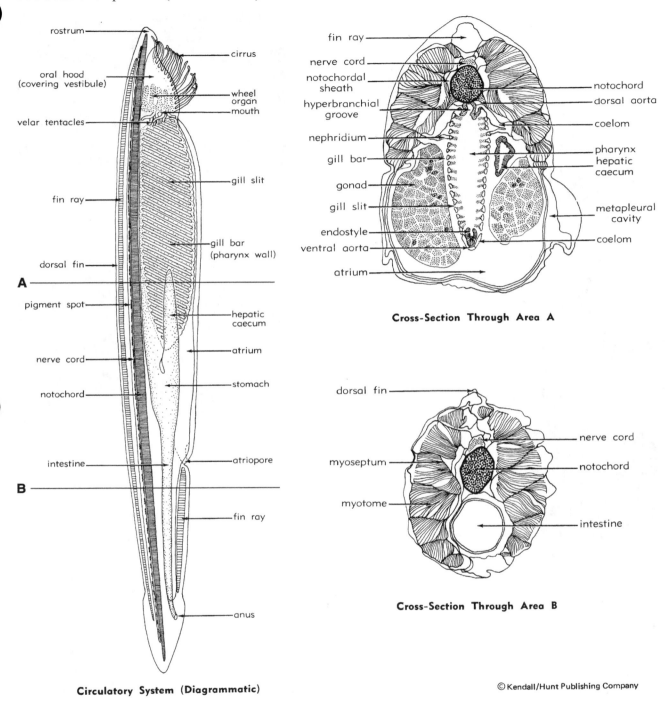

Circulatory System (Diagrammatic)

Cross-Section Through Area A

Cross-Section Through Area B

© Kendall/Hunt Publishing Company

feature to look for are the slitted **gill** openings behind the mouth which are uncovered. Touch the skin of a preserved shark specimen and note its abrasiveness due to the enamel-like **placoid scales** (Plate 9).

Class Osteichthyes

The bony fish are represented by familiar hobby, game and food varieties such as; perch, tuna, guppy, catfish, salmon, trout and eel. Look at the various fish specimens on the demonstration table and point out the paired **fins, homocercal tail, terminal mouth** and bony **operculum** covering the gills. If a slide is

PLATE 8 Agnatha (Lamprey).

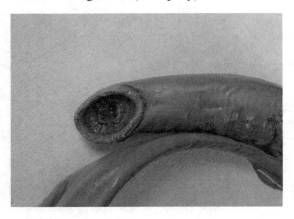

PLATE 9 Chondrichthyes (Ray).

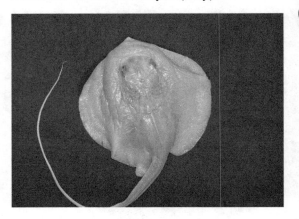

PLATE 10 Osteichthyes (Sea Horse).

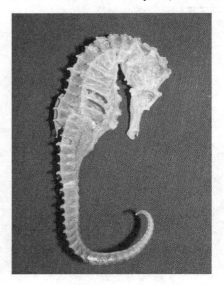

PLATE 11 Amphibia (Salamander).

available, observe the differences between the types of scales: **ganoid, cycloid** and **ctenoid.** If a fish skeleton has been cleaned and mounted, look at the skeletal system composed of **bone** (Plate 10).

Class Amphibia

When adapting to terrestrial existence, amphibians modified their body for: air breathing, prevention of dessication, fluctuating temperatures, and movement on land. They still retain their connection to a watery past because of their mode of reproduction. Eggs are shed in aquatic environments and the young generally pass through a gill breathing, swimming, **tadpole** stage before converting to limbed, air-breathing land dwellers. Most adult amphibians have four limbs (**tetrapods**) although the **caecilians** have a slender worm-like body. Their skin is **moist,** often with poison glands, and their head and trunk are fused with no neck region. **Salamanders** and **newts** are carnivorous amphibians that possess a tail. Observe live or preserved varieties such as *Ambystoma* (axolotl) or *Necturus* (mudpuppy). In *Necturus,* the sexually mature adult retains the larval gill features; a condition called **paedomorphosis** (Plate 11).

PLATE 12 Reptilia (Turtle).

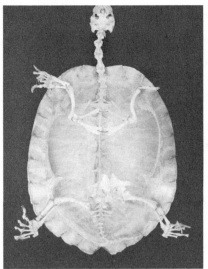

PLATE 13 Aves (Bird).

PLATE 14 Mammalia (Skull).

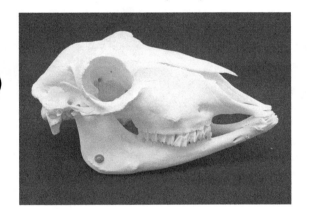

Frogs and toads have specialized hind limbs for jumping and leaping. They pass through a transformation that absorbs the tail, grows limbs and develops internal lungs. The leopard frog, *Rana pipiens,* is the typical amphibian used in biology laboratories and will be studied thoroughly in the next exercises.

Class Reptilia

The body of reptiles is covered with thickened epidermal **scales.** Members are lung-breathers and may be limbless or tetrapods. They are freed from the aquatic dependence for reproduction because they produce a calcareous or leathery, **shelled egg.** Four extraembryonic membranes: the **chorion, amnion, allantois** and **yolk sac,** protect the developing embryo and surround it in a watery environment. Look at the scales of limbless snakes and compare the skin to that of the shelled turtles. Lizards, geckos, crocodiles and alligators all possess the common reptilian features (Plate 12).

Class Aves

Bird members are easily identified by their body covering of **feathers.** Although flight is not possible in all birds, they have **wings** as forelimbs and **scaled legs.** Look at several types of feathers. Observe one under a dissecting microscope. Ascertain the habitat and diet of different birds on the display table by comparing their beaks and feet. Look for the **keeled** sternum and long neck on a pigeon skeleton. Compare the size, shape and color of various bird eggs (Plate 13).

Class Mammalia

All mammals possess **hair, mammary glands,** movable **eyelids** and **fleshy ears.** The mouth, with its **dentition formula,** is a means of identification in mammalian orders. Check out various **skulls** and look for differences in their teeth shape and numbers. If taxidermy preparations of skins are available, be able to recognize some common members. Compare the skeleton of a cat to a bird, snake, turtle, frog and fish (Plate 14).

Questions

1. Of the vertebrate members, which groups are predominantly aquatic? Predominantly terrestrial?

2. Describe the major problems for multiple systems when changing from life in the water to a terrestrial existence.

3. Which of the chordate members produces an amniotic egg?

4. Name the specific three groups (invertebrate or vertebrate) of animals who have achieved true flight.

5. What is the dentition formula for a human?

6. What type of fertilization, external or internal, is found in reptiles and birds with shelled eggs?

7. What is the economic significance of each of the vertebrate groups?

EXERCISE 15

Embryology: Comparative Morphologies and Strategies of Development

Objectives

After completion of this exercise, the student should be able to do each of the following:

Identify the major stages of starfish development under the microscope.

Define fertilization, cleavage, blastula, gastrula, and morphogenesis.

Identify the major stages of frog and chick development under the microscope.

List the differences in the four types of embryonic development discussed.

Explain the function of each of the four extra embryonic membranes.

List the three germ layers and name at least two structures derived from each.

Answer the review questions at the end of this exercise.

INTRODUCTION TO EMBRYOLOGY

In the process of sexual reproduction in animals, two gametes—sperm and egg—fuse to make one new cell. These two gametes are usually donated by two different parents. The new zygote resulting from this fusion contains material from both parents and is "new" in the sense that it contains new potentialities resulting from the mixture of the genetic material from the two parents. Embryology is the study of the development of the zygote, the product of sexual reproduction.

The zygote develops as a result of three kinds of activity: **mitosis** and subsequent growth, **differentiation of cells,** and **movement of cells.** Thus, as a result of embryonic activity, the one-celled fertilized egg (zygote) changes or develops into the adult form eventually. In observing the embryonic development of any animal we can divide it into the following stages:

1. **Fertilization**—the fusion of nuclei and other events associated with the union of sperm and egg. The result is the zygote.
2. **Cleavage**—a series of mitotic divisions undergone by the zygote. No growth in size of the entire structure occurs at this time. The actual pattern of cleavage—whether the whole zygote **(holoblastic cleavage)** or only a part of it divides **(meroblastic cleavage)**—varies from one species of animal to another, depending on the amount of yolk in the egg and its distribution. The end result of cleavage is the formation of a hollow ball of cells known as a **blastula,** the central cavity of which is known as the **blastocoel.**

3. **Gastrulation**—the migration of cells of the blastula resulting in the formation of a new cavity known as the **gastrocoel** (archenteron, or "Primitive gut"). Further development leads to the formation of three distinct **germ layers** (embryonic cell layers) known as the **ectoderm, mesoderm,** and **endoderm,** from which all the organs of the new organism develop. At this stage, when the three germ layers are present, the structure is technically termed an embryo.

The development of internal shape characteristics of the animal, known as **morphogenesis,** results from the shaping of the germ layers due to differential growth, movement, and association of cells in the germ layers.

4. **Neurulation**—the development of the notochord, neural tube, and coelom. This only occurs in chordates.

5. **Organogenesis**—the differentiation and association of cells to form organ systems.

In this exercise, we will make a brief survey of four different animals: starfish, frog, chick, and human. This survey should, first of all, illustrate the basic similarities in the embryonic development of these entirely different animals. Secondly, even though there are similarities, there are also variations, based primarily on the amount of yolk present in the cytoplasm of the egg.

STARFISH DEVELOPMENT

Obtain one slide of the starfish development. On this slide, you will find all of the stages indicated in Figures 1–7. The stages on the slides are whole mounts, not sections. Identify each of the stages indicated on the slide.

NOTE:
Draw representative stages of starfish development as indicated on Figure 8.

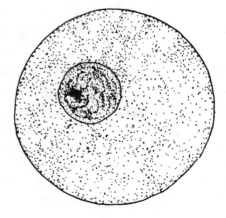

FIGURE 1 Unfertilized Egg.

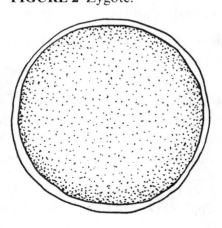

FIGURE 2 Zygote.

Unfertilized Egg

In this nearly spherical cell a large nucleus and a nucleolus are clearly visible. A small amount of yolk (stored food) is present in the form of many small particles. Are these yolk particles present in any particular area of the cytoplasm, or are they scattered about?

Zygote (Fertilized Egg)

This is a single celled structure very much like the unfertilized egg in appearance. In contrast to the unfertilized egg, the zygote's nucleus is inconspicuous. Would this structure be 1N or 2N?

Early Cleavage

The 2-, 4-, and 8-celled stages are included in early cleavage (Figure 3). Find an example of each of these and note that the cells remain attached to each other. Is there any growth in size?

How does the cell size in each of these stages compare with the size of the zygote?

Are all the cells within a single stage the same size?

FIGURE 3 Early Cleavage.

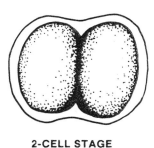 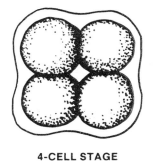 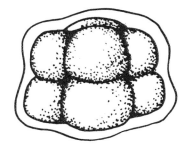

2-CELL STAGE   4-CELL STAGE   8-CELL STAGE

FIGURE 4 Later Cleavage.

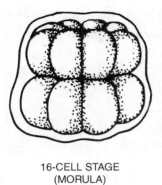 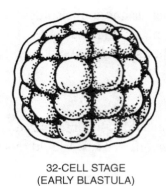 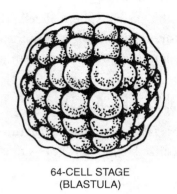

16-CELL STAGE
(MORULA)

32-CELL STAGE
(EARLY BLASTULA)

64-CELL STAGE
(BLASTULA)

Later Cleavage

The 16-, 32-, and 64-celled stages are included in later cleavage. The 64-celled stage is hollow and is called the **blastula.**

What happened to the individual cell size?

Blastula

As cell division continues, the increasing number of cells become arranged around an enlarging central cavity known as the **blastocoel** (Figure 5). In the starfish, the walls of the blastula are usually one cell layer thick. Why would you expect your specimen to appear dark around the edges and light in the middle?

FIGURE 5 Late Blastula Cross-Section.

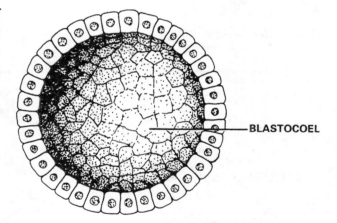

BLASTOCOEL

Can you see any differences among the cells?

How does the size of the blastula compare to earlier stages?

Gastrula

Shortly after the formation of the blastula, a small depression begins to appear at one end of the embryo; the appearance of this depression marks the beginning of gastrulation. As gastrulation proceeds, the depression invaginates (folds inward) more and more. Which of the three basic embryonic activities would this be?

Find the embryos at various levels of gastrulation on your slide. The latest stages on your slide are those in which the inner end of the invagination is beginning to expand.

As a result of gastrulation in the starfish, the embryo produces two primary cell layers: an outer **ecto-derm,** and an inner **endoderm.** The third layer, the **mesoderm,** develops later, between these two. Gastrulation, then, eventually results in an embryo with 3 primary germ layers, a mere remnant of the blastocoel, and a new cavity known as the **gastrocoel** or **archenteron.** The gastrocoel is continuous with the outside through the **blastopore** and which will become the cavity of the digestive tract. In the deuterostomes, the blastopore will eventually become the anus.

FIGURE 6 Gastrulation.

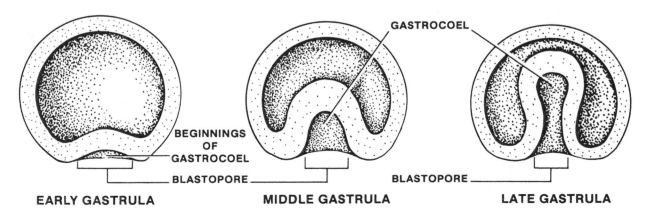

EARLY GASTRULA   MIDDLE GASTRULA   LATE GASTRULA

Larval Stage

The gastrula stage is reached within a day or two after fertilization. Within another day or two, this stage undergoes some alterations to give rise to the **larval stage,** which is free-swimming. During a period of from several weeks to several months, the larva grows. After this time, it settles to the bottom and becomes a small starfish. Refer to Figure 7.

FROG DEVELOPMENT

Examine the charts, models, and whole specimens available in lab. Obtain slides of sectioned embryonic stages. Using Figures 9 through 16, locate the structures printed in boldface in this part of the exercise.

Unfertilized Egg

The egg consists of two portions: a darkly pigmented portion, the **animal pole,** and a lightly colored, yolk-filled portion, the **vegetal pole.** In nature, when the egg is

FIGURE 7 Bilateral Starfish Larva.

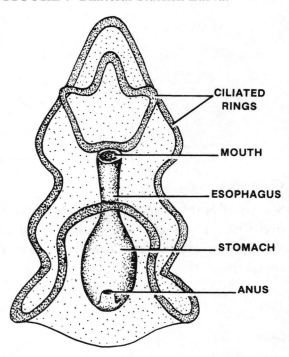

CILIATED RINGS

MOUTH

ESOPHAGUS

STOMACH

ANUS

released into the water, the **gelatinous covering** produced by the oviduct absorbs water and swells, causing the eggs to be equidistant from each other. What is the importance of this swelling?

Zygote

Sperm have to penetrate the eggs before swelling of the gelatinous covering takes place. Once fertilization has occurred, the cell rotates in such a way that the heavier portion of the cell, the yolk-filled vegetal pole, is downward.

Suppose you had a dish of fertilized frog eggs; how would you know if they were all fertilized (i.e., what color would be facing upward)?

FIGURE 8 Stages of Starfish Development.

UNFERTILIZED EGG   FERTILIZED EGG   TWO-CELL STAGE

FOUR-CELL STAGE   EIGHT-CELL STAGE   MORULA

BLASTULA, SECTION   EARLY GASTRULA, SECTION   MIDDLE GASTRULA, SECTION

LATE GASTRULA

FIGURE 9 Frog Zygote.

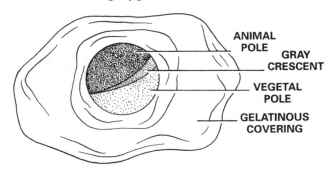

ANIMAL
POLE

GRAY
CRESCENT

VEGETAL
POLE

GELATINOUS
COVERING

Another indication that fertilization has occurred is the appearance of a pigmented area (**gray crescent**) between the yolk-filled and black portions.

Cleavage

The beginning of the first cleavage is marked by the appearance of a groove on the animal pole end of the egg; this **cleavage furrow** gradually extends toward the opposite side of the zygote, dividing it into two cells. The next cleavage occurs at right angles to the first and produces the **4-cell stage.** The third cleavage occurs parallel to, but a little above the equator of the developing embryo. What would you suspect was the cause of this unequal division?

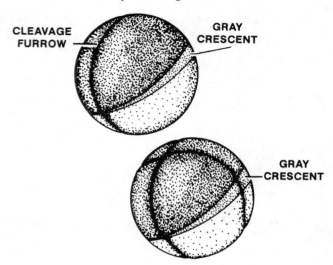

FIGURE 10 Early Cleavage.

In which region of the developing embryo would you expect a more rapid rate of cell division?

What effect would this have on cell size in the two regions of the developing embryo?

Figures 11 and 12 are two diagrams of early and late cleavage. Note that the jelly layer is not shown in these and subsequent stages. In the space below each of the drawings, make a diagram of the section through these stages as seen in the prepared slides. Label the **animal pole, vegetal pole,** and **cleavage grooves** on your drawings. Can you find the yolk particles?

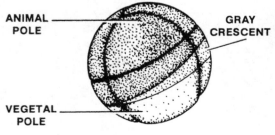

FIGURE 11 Third Cleavage.

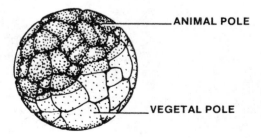

FIGURE 12 Late Cleavage.

Give two ways in which the animal pole can be distinguished from the vegetal pole.

Is the **pigment** localized in any particular region of the cell cytoplasm or is it distributed equally throughout?

What do you think is the function of the **membrane** (blue-purple color) around the entire structure on the slides?

Blastula

The stage in which the **blastocoel** (internal cavity) is formed is called the blastula. Notice that, instead of being centrally located as in the starfish, the blastocoel is off center toward one pole of the developing embryo.

In which hemisphere is it located? Why?

Gastrula (Yolk Plug Stage)

The large amount of yolk in the frog egg prevents the type of invagination at one end of the blastula which is seen in the starfish. Instead, the more rapidly dividing cells from the animal pole grow downward over the yolk-filled cells, gradually enclosing them. **Gastrulation** begins with the pushing inward (or **invagination**) of these cells. This forms a crescent-shaped "line" on the surface known as the **dorsal lip.** The opening marked by the dorsal lip is known as the **blastopore.** This marks the posterior end of the embryo. In the prepared slide of the early gastrula, you can actually see a depression forming here as the surface cells move inward. The depression gradually enlarges to form the **gastrocoel** or archenteron, as seen in the slide of the late gastrula. Note the difference between this gastrocoel and the one in the starfish.

FIGURE 13 Frog Blastula.

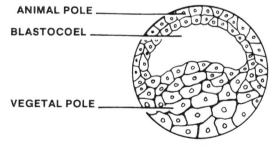

LONGITUDINAL SECTION, BLASTULA

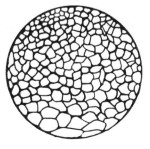

INTACT LATE BLASTULA

FIGURE 14 Frog Gastrulation.

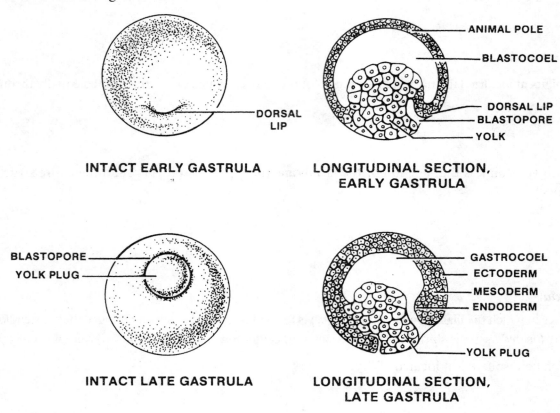

What will the gastrocoel give rise to?

With continued cell division and the migration of cells inward, the animal pole cells encircle, and eventually line, the entire gastrocoel except for a small circular plug of yolk cells within the blastopore opening,
the **yolk plug.** Eventually, even the yolk plug disappears as it is covered by the migrating cells. As all of
this is occurring, the three developing **germ layers** are becoming arranged in such a way that there is an
outside layer of cells (**ectoderm**), an inner lining of the gastrocoel (**endoderm**), and a layer between the
two (**mesoderm**).

Can you detect any visible differences between the cells of these three layers?

Is there any difference in the overall shape of the structure?

Neurula

Near the end of gastrulation, the ectodermal cells in the mid-dorsal region of the embryo thicken to form a flattened area on the surface known as the **neural plate.** The sides of the neural plate, the **neural folds,** gradually fold upward forming a depression, the **neural groove,** between them. Eventually the folds fuse, forming a closed tube, the **neural tube,** which will develop into the brain and spinal cord. Refer to Figure 15.

In the midline of the mesodermal layer, the cells develop into a cylindrical rod, the **notochord.**

FIGURE 15 Neurulation.

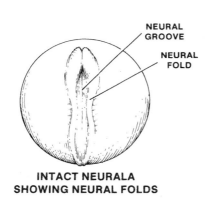

**INTACT NEURALA
SHOWING NEURAL FOLDS**

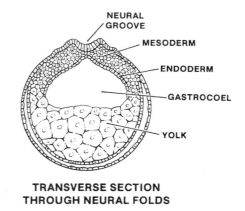

**TRANSVERSE SECTION
THROUGH NEURAL FOLDS**

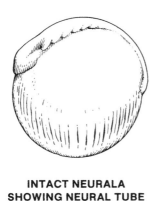

**INTACT NEURALA
SHOWING NEURAL TUBE**

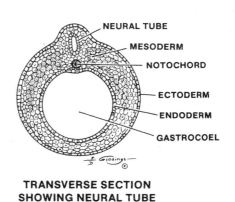

**TRANSVERSE SECTION
SHOWING NEURAL TUBE**

Although the notochord is a rod extending along the length of the animal, how would you expect it to appear in your slide?

FIGURE 16 Frog Larval Stages.

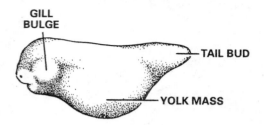

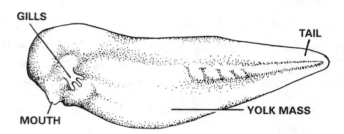

What is the cavity in the area below the notochord?

What is the function of the large cells that may be (depending upon where the animal was sectioned) located in this region?

Larval Stage

With the formation of the neural tube, the embryo elongates. The anterior end becomes slightly enlarged, forming the **head.** A tail develops from the posterior end, and **gills** develop for gas exchange. The mouth opens so that feeding can occur. Over a period of growth of a few months, this **tadpole** will metamorphose into an adult frog capable of survival both on land and in water.

CHICK DEVELOPMENT

In both the starfish and the frog, all of the cells derived from the fertilized egg are used in making the new individual. In land animals, however, a number of cells are not used in the immediate makeup of the embryo body; instead, they give rise to certain temporary structures necessary for embryonic development on land. Figure 17 demonstrates the four membranes that develop in the chick to ensure survival:

FIGURE 17 Chick Embryonic Membranes.

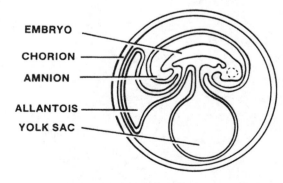

 Amnion—contains water to shield the embryo.

 Yolk sac—contains yolk as a food source.

 Allantois—stores waste materials produced by the developing embryo plus transporting respiratory gases between the embryo and its environment.

 Chorion—also functions in exchange of gases.

Thus, the egg of the chick is a "self-contained" capsule simulating the watery environment of the lower vertebrates and yet showing remarkable adaptations to land development.

Unfertilized Egg

Examine the egg on demonstration in lab. Figure 18 shows the internal structure of the chicken egg. Notice that in this case the everyday term of "egg" refers to more than simply the egg cell, **ovum.** The circular yellow mass that we would refer to as "yolk" as we are sitting at the breakfast table is actually the ovum. The yolk of the egg is in fact the single egg cell. It is composed of a large amount of **yolk granules** plus a small amount of yolk-free cytoplasm, the **germinal disc,** on the surface of the yolk. It is the germinal disc that contains the nucleus to be fertilized and that will undergo cell division to form the embryo. If the egg becomes fertilized, the germinal disc is referred to as the **blastoderm.** The yolk supplies nourishment to the developing embryo.

After the embryo has been ovulated, it will pass down the oviduct acquiring a number of other structures designed to aid survival in the land environment. Immediately around the ovum is the egg "white"; this is known as **albumen,** which functions as a water reservoir for the young embryo and as a food source for later development. Attached to the yolk and extending out into the albumen are dense cordlike structures, the **chalazae,** that serve to suspend the yolk in the albumen. In another region of the oviduct, two thin shell membranes are deposited around the albumen; these function in decreasing water loss. The next section of the oviduct produces the **shell** and molds the egg into its customary shape. The hard protective shell is porous, allowing for gas exchange.

Early Development

By the time the egg is laid, the stages of development through gastrulation have usually occurred, and the embryo is undergoing neurulation. Notice that only the blastoderm undergoes cleavage. Note also that, because of the tremendous amount of yolk, the blastula is not a "hollow ball of cells"; instead, it is a hollowed, flattened disc on the yolk surface. A third difference is with regard to gastrulation. Notice that the blastopore (the site where surface cells migrate inward to establish the three germ layers) is not

FIGURE 18 Internal Structures of Hen's Egg.

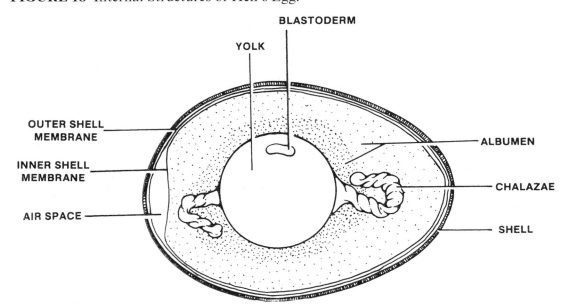

FIGURE 19 Stages in the Development of the Chick.

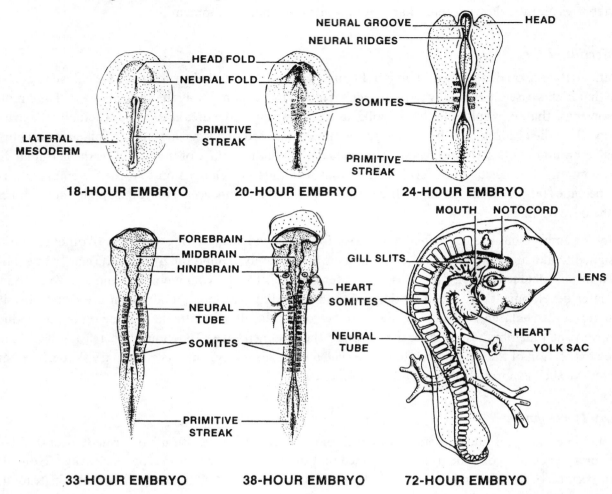

a circular opening as was in the case of the starfish and the frog. In the chick, the blastopore is elongate and is referred to as a **primitive streak.**

Later Development

Obtain plastomounts of different stages of chick development and examine them under a stereomicroscope for the structures in Figure 19. Notice particularly the characteristic curvature of the embryo body as it is being separated from the underlying yolk. The circulatory system within the embryo is quite extensively developed; the heart would definitely be beating if this were alive. Why do you suppose there is such an extensive blood vessel network outside the embryo body? The 72-hour embryo has two extreme bends which make it curve back on itself. Identify the **anterior limb bud** which will give rise to the wing.

Notice the bump-like structures on either side of the spinal cord (mid-dorsal region). These structures are the **somites,** blocks of mesodermal cells that will later give rise to muscles and vertebrae. They are also used as a means of determining how old the embryo is. How many pairs are there.

In looking at the embryo, can you determine if there is a difference in the rate of development in comparing the anterior and posterior regions of the body? Explain your answer.

Chick development continues for 21 days, until hatching. Note that this animal does not go through a larval stage as did the starfish and frog.

EXTRAEMBRYONIC MEMBRANES OF THE CHICKEN AND MAMMALS

Human development parallels that of the chick even though the egg of humans does not contain an overabundance of yolk. Why?

The reptiles were the first to lay eggs on land. Their eggs contained extraembryonic membranes by which the embryo carried out gas exchange, excretion of wastes, and consumption of stored food (yolk). These same membranes develop in the human, but are put to different uses since the human develops internally. Figure 20 shows the membranes in a chick egg and compares them to those in the human.

FIGURE 20 Comparison of the Extraembryonic Membranes in the Human and Chick.

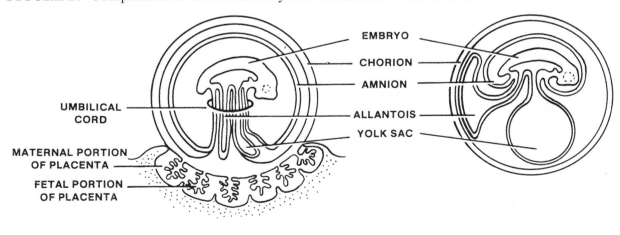

HUMAN DEVELOPMENT

Referring to Figure 21, note that fertilization normally occurs in the upper 1/3 of the oviduct, and early development of the embryo occurs during its movement down the oviduct. By the time the developing embryo reaches the uterine cavity (5–7 days after ovulation), it is in the **blastocyst** stage.

The blastocyst is equivalent to the blastula stage; the blastocyst cavity does not have a wall of uniform thickness: a cluster of cells occurs at one end. The wall of the blastocyst will give rise to the **chorion.** This membrane has two important functions: (1) during early development (first 3 months) it produces **HCG,** the hormone responsible for keeping the corpus luteum functioning; and (2) it initially absorbs nutrients from the endometrium through villi and later gives rise to the fetal portion of the **placenta.** The cluster of cells is known as the **inner cell mass,** which will give rise to 4 structures; **embryo, amnion,** which forms a protective, fluidfilled cavity around the developing embryo; **yolk sac,** present even though the human

FIGURE 21 Diagrammatic Representation of the Movement of the Oocyte and Embryo.

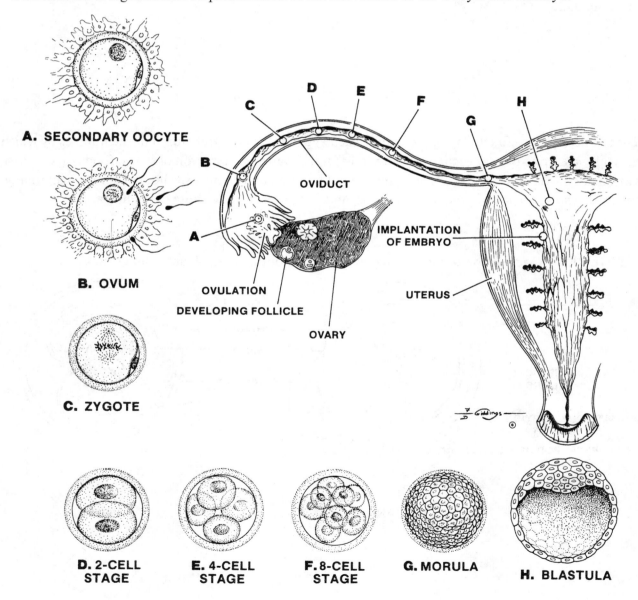

oocyte has no yolk granules; and **allantois,** whose most important function is probably the formation of the umbilical blood vessels. (Refer to Figure 20 for the structures printed in boldface.)

A few days after entering the uterus, the blastocyst undergoes the process of **nidation** (implantation), during which it buries itself in the endometrium from which it will acquire nutrients for the rest of its development.

Gastrulation occurs in somewhat the same manner as in the chick since a primitive streak is formed. The **embryo** with its 3 germ layers (ectoderm, endoderm, and mesoderm) is formed within 2 weeks after ovulation. By the end of 8 weeks of development, all of the body parts are present and it is now called a **fetus,** rather than an embryo. After this time, the organs develop further to eventually become functional and the fetus grows rapidly in size. To accommodate the increased nutritional demands for this growth, the **placenta** serves as the site of nutrient and waste exchange between fetus and mother.

To give you a better appreciation of the growth involved, the following chart indicates the "crown-rump" length (i.e., sitting height) of the human fetus at different stages of development:

    2 weeks—0.23 mm
    1 month—3/8 in.
    2 month—1 in.
    3 month—3 in.
    4 month—5 in.
    5 month—7 in.
    6 month—9 in.
    7 month—11 in.
    8 month—14 in.

Examine the materials on display in the lab. The models display the relationship of the fetus to the uterus, and the charts illustrate the process of birth.

Fetal Pig Dissection

Objectives

After completion of this exercise, the student should be able to do each of the following:

Locate and identify the anatomical parts of the fetal pig given in boldface in this exercise, and give the function of each.

Define the terms of relative position and direction given in the introduction, and locate these positions and directions on any vertebrate.

Distinguish between male and female fetal pigs on the basis of their external features.

Locate the parts of the respiratory system seen in the fetal pig on the human torso model and color plates in the laboratory manual.

Locate as many as possible of the parts of the digestive system given in boldface in this exercise on the charts and models which show the human digestive system.

Locate the parts of the heart given in boldface in this exercise on the sheep heart and on the human torso model.

Answer the review questions at the end of this exercise.

INTRODUCTION

In this exercise, the anatomy of the fetal pig, *Sus scrofa*, will be studied. This will aid us in obtaining some knowledge of mammalian anatomy and physiology. You will also examine certain comparative aspects of the anatomy and physiology of man.

The animals that you will use were removed from their mothers which were slaughtered for food. Farmers market pregnant females because hogs are sold by the pound.

In the pig, fertilization of the egg occurs in the oviduct; by the time the early embryo reaches what is equivalent to the blastula, it becomes buried in the uterine wall where subsequent development takes place. Here not only do the extraembryonic membranes form, but also a new organ, the **placenta**. This organ is made up of the **extraembryonic membranes**: amnion, chorion, and the allantois, which become intricately intermingled with the lining of the uterus. Remember that at no time does the blood of the fetus (unborn) mix with that of the mother. Gases and small molecules are able to pass across capillary walls in the placenta. The fetus lives like a parasite on the mother, absorbing all of its nourishment and

From *Encounter with Life: General Biology Lab Manual,* ISBN 0-89582-252-0 by Hans F.E. Wachtmeister and Larry J. Scott. Copyright © 2006 by Morton Publishing Company. Reprinted by permission.

oxygen from and excreting all of its wastes into the blood of the mother, all by way of placental circulation. The connection between the placenta and the fetus is through the **umbilical cord**. The period or length of pregnancy (gestation) in pigs is approximately 17 weeks. The fetal pigs used in lab will be within one or two weeks of birth.

Below is a list of anatomical terms which are used frequently in dissection directions. Familiarize yourself with the definitions before continuing this exercise.

Dorsal—near or toward the back.

Ventral—near or toward the belly.

Medial—near or toward the middle.

Lateral—near or toward the sides.

Anterior—near or toward the head end.

Posterior—near or toward the tail end.

Caudal—referring to the tail or tail end.

Cephalic—referring to the head or head end.

Longitudinal—in the axis from head to tail.

Transverse—a thin section which cuts across the body at a right angle to the long axis.

Superficial—on or near the surface.

Pectoral—relating to the chest or shoulder region.

Pelvic—relating to the hip region.

Distal—free end of a limb or projection or toward this free end of a limb or projection.

Proximal—end attached to the body, or toward the end attached to the body.

The fetal pigs have been preserved in either formalin or isopropyl alcohol. These preservatives tend to dehydrate your fingertips; therefore, students may desire to lightly grease their fingertips with petroleum jelly.

The instructor will make available one fetal pig for each two students. The fetal pigs will be evenly distributed in reference to male and female. Upon receiving a fetal pig, wash it in running water and place it in a dissecting pan for observation. The aborted pigs will usually have a slash on one side of the neck. This marks the location where the blood was drained from the pig and where red and blue liquid latex were injected into a major artery or vein, respectively. The latex has become solid and rubbery in texture in the fetal pig. This strengths the blood vessels and aids in their identification.

Measure your pig from snout to anus and refer to the following table to determine its approximate age. Remember that 1 cm. = 10 mm.

11 mm.—21 days

17 mm.—35 days

28 mm.—49 days

40 mm.—56 days

220 mm.—100 days

300 mm.—115 days (full term)

Please bear in mind, as you dissect your fetal pig, that there are very few differences between the anatomy of the pig and that of the human being. As you dissect out each system, make a mental comparison between yourself and the fetal pig concerning the arrangement of the internal organs.

EXTERNAL ANATOMY

Beginning at the anterior end, locate the **mouth**, which leads into the **oral cavity**, the **nose**, the **nostrils**, which lead into the **nasal cavity**, the **ears**, the **external ear canal**, which leads inward from the ears. Locate the **nictitating membrane** in the corner of the eye, the **eyeball, eyelashes**, and **eyelids**.

Now, lay the pig on its side and identify the major body divisions beginning anteriorly with a large **head**, a short thick **neck**, a cylindrical **trunk** with two pairs of **appendages**, and a short **tail**. The **anus** is located ventral to the tail.

Examine the forelimbs and locate the **shoulder, elbow**, and **foot**. On the hindlimbs find the **hip, knee, hock joint** (ankle), and the **foot**.

Turn the pig ventral side up and locate the large **umbilical cord** in the abdominal region. This cord connects the fetus to the mother at the placenta. Cut a half inch off the umbilical cord and observe the umbilical blood vessels. The blood vessels in the umbilical cord consist of two small **umbilical arteries** having relatively thick walls and an **umbilical vein**, considerably larger than the two arteries. The umbilical arteries carry blood from the fetal pig to the placenta, while the single umbilical vein returns blood to the fetal pig's body from the placenta. A fourth very small vessel, the **allantoic duct**, serves to carry some of the small amount of urine formed by the kidneys away from the fetus. Refer to Figure 1. Make a sketch below of a section through the umbilical cord, and label the four vessels.

Also, on the ventral surface on either side of the umbilical cord, find the row of **nipples** or **teats**. The number of nipples indicates the number of mammary glands.

Next, determine the sex of your pig. You will be expected to recognize both male and female pigs, even though you have only one sex for dissection. In the male, the opening of the urogenital tract, which serves both reproductive and excretory functions, lies at the end of the **penis**, just posterior to the umbilical cord. The **scrotum**, which contains the **testes**, is situated ventrally with respect to the anus. The female pig has a single urogenital opening, located ventral to the anus, with a dorsal projection of tissue, **genital papilla** below it.

DISSECTION OF THE FETAL PIG

For this dissection you will need string, scissors, a sharp scalpel or a single edge razor blade, a blunt probe, forceps, and some dissecting pins. Place the pig on its dorsal surface in the dissecting pan. Referring to Figure 1, tie a string around one forelimb and place the string around and under the pan and tie it to the other forelimb. Make sure that the forelimbs are spread apart. Repeat this procedure for the hindlimbs. Note: When preparing to put the pig back in the storage container for future use and reference, do not untie the string. Simply slide the pan from underneath the pig. Also, put a tag on the pig, identifying the members of the group. Use a pencil when writing on the tag to avoid fading.

Dissecting does not mean cutting up the specimen. Instead, it means exposing a specimen to view. Use the scalpel or razor blade carefully and sparingly. You may find that the most useful tool is the dull

FIGURE 1 The Fetal Pig (External Features and Dissection Guidelines)

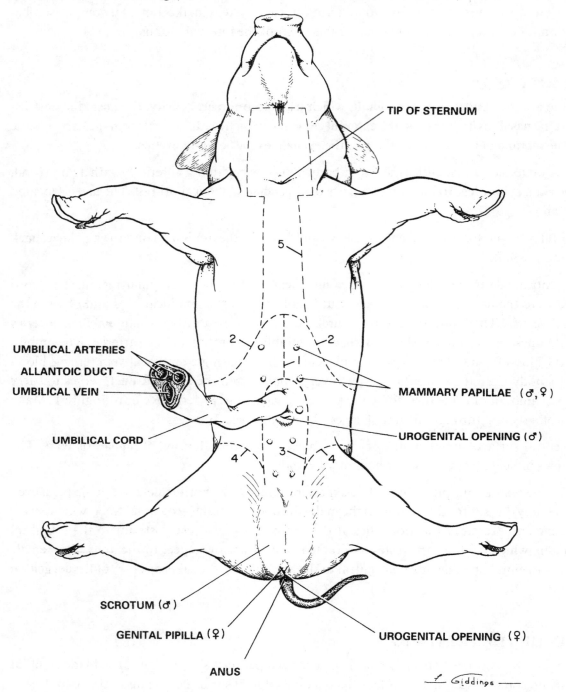

TIP OF STERNUM

UMBILICAL ARTERIES

ALLANTOIC DUCT

UMBILICAL VEIN

UMBILICAL CORD

MAMMARY PAPILLAE (♂,♀)

UROGENITAL OPENING (♂)

SCROTUM (♂)

GENITAL PIPILLA (♀)

UROGENITAL OPENING (♀)

ANUS

probe, which can be used to separate organs from membranes. Be very careful with your dissection. Any tissues or organs that are removed should be deposited in a trash can and never in the lab sink.

Respiratory System

This system functions in the exchange of gases between the internal and external environments of the organism. Air enters and leaves through the **nostrils** which lead, via the nasal cavity, into the **pharynx** or throat. The pharynx is where food and air passages cross and is located posterior to the **oral cavity**. Expose the pharynx by inserting scissors into the corners of the mouth and cutting the jawbones. You may have to cut approximately one to one and a half inches. Separate the jaws further by pushing down on the tongue until you find a cartilaginous projection, the **epiglottis** at the base of the tongue. This flap aids in preventing food from entering the air passageway which leads into the lungs. The epiglottis covers an opening known as the **glottis**, the opening of the windpipe or trachea. Refer to Figure 3-A.

Figure 1 indicates the incisions required to open the thoracic and abdominal regions. A scalpel and forceps should be used during this part of the dissection. Make cuts as indicated by the dotted lines in Figure 1. To enter the abdominal cavity, you will pass through an outer layer of skin, several layers of muscle, and a tough inner glistening membrane, the **peritoneum**, which lines the body cavity and surrounds the internal organs.

Pick up the umbilical cord and pull lightly on it. Passing from the base of the umbilical cord to the liver, the large brown organ covering most of the anterior portion of the abdominal cavity, is the **umbilical vein**. After identifying it, cut it, leaving two stump ends that can be located later.

A certain amount of brown liquid may be present in the body cavity. This is clotted blood and it should be poured into the sink, after which you should completely rinse the body cavity with cold tap water. Now, pin back the skinflaps, and fold back between the hind legs the skin and muscle to which the umbilical cord is attached.

Locate the muscular **diaphragm** which separates the thoracic cavity from the abdominal cavity, and aids in the movement of gases into and out of the lungs.

To expose the viscera in the thoracic region, continue the initial incision from the abdominal cavity forward to the clump of hairs under the chin. Gradually deepen the incision until you have cut through the **sternum**, or **breastbone**. The muscle is particularly thick in the neck region. Avoid cutting the organs and blood vessels in the neck and chest regions. Sever the edges of the diaphragm where it is attached to the rib cage. Next, break the rib cage by applying pressure from your thumb on the sternum. The flap of tissue which contains the sternum can be carefully trimmed away and discarded. Some of the ribs may have to be cut with scissors. If so, be careful not to cut into the organs of the thorax.

Referring to Figure 2 locate the **larynx**, or voicebox, and slit it open along its midline to expose the small, paired lateral flaps internally known as the **vocal folds**. In the fetal pig, these are not yet well-developed. Posterior to the larynx is the cartilaginous-ringed **trachea**. In the neck you will expose extensions of the **thymus gland**. The major portion of this gland is located on the ventral surface of the anterior portion of the heart. The extensions of the thymus gland may have to be removed. Also, in a midventral position, locate the small, dark, pea-shaped **thyroid gland** which lies on the surface of the trachea just anterior to the heart. Leave the thyroid gland in place.

FIGURE 2 The Fetal Pig (Respiratory System)

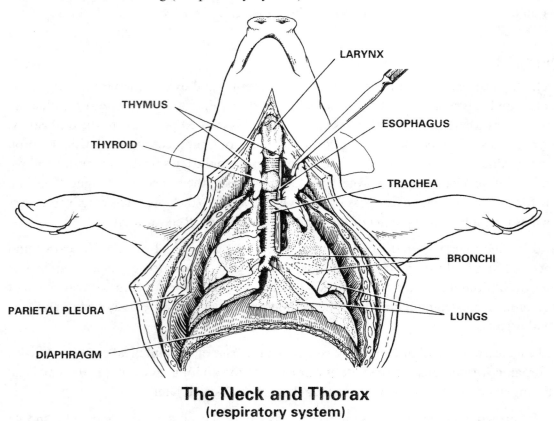

LARYNX

THYMUS

THYROID

ESOPHAGUS

TRACHEA

BRONCHI

PARIETAL PLEURA

LUNGS

DIAPHRAGM

The Neck and Thorax
(respiratory system)

Follow the trachea posteriorly until it branches into two **bronchi**, one leading to each **lung**. The left and right lungs are in separate cavities lined by a thin, fleshy **pleural membrane**. Trace the path of a bronchus into a lung by scraping away the lung tissue. The bronchus continues to branch into smaller tubes called **bronchioles** which lead to tiny air sacs or **alveoli** which are surrounded by a network of blood vessels. The alveolus is the site of gas exchange in the lungs.

In order to observe the nature of the lung tissue, remove the anterior lobe of the left lung. It should be noted that all dissections regarding the pig refer to the body of the pig. Therefore, if the directions identify an organ on the left side, this refers to the left side of the pig. Place the lung in a small dish of water. Holding it with your forceps, gently tease the lung apart with the blunt wooden base of a probe. Using a stereomicroscope, identify the branching network of bronchioles and blood vessels.

Having examined the respiratory system in the pig, consult the human torso model and the color plates of the human respiratory system in the laboratory manual. Identify the main structures of this system.

Digestive System

Digestion begins in the **oral cavity** with the chewing of food. The **tongue** is attached at the back of the oral cavity. Observe the small, underdeveloped **teeth** in both the upper and lower jaws. The first set of teeth in mammals is called **milk teeth**. These are later replaced by the **permanent teeth**. Referring to Figure 3 note the ridged surface on the roof of the oral cavity. This is the **hard palate**. Posterior to the hard palate is the **soft palate**.

FIGURE 3 The Fetal Pig (Digestive System)

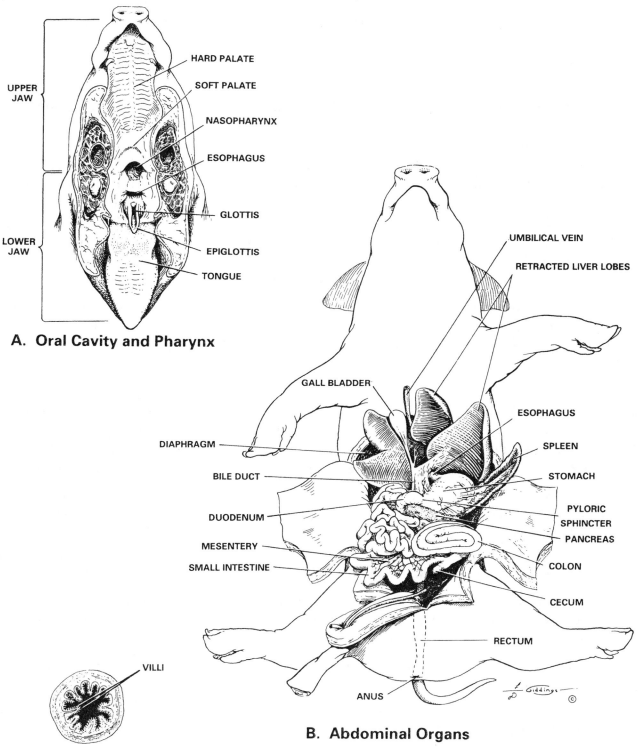

HARD PALATE
SOFT PALATE
NASOPHARYNX
ESOPHAGUS
GLOTTIS
EPIGLOTTIS
TONGUE

UPPER JAW

LOWER JAW

A. Oral Cavity and Pharynx

UMBILICAL VEIN
RETRACTED LIVER LOBES
GALL BLADDER
DIAPHRAGM
BILE DUCT
DUODENUM
MESENTERY
SMALL INTESTINE
ESOPHAGUS
SPLEEN
STOMACH
PYLORIC SPHINCTER
PANCREAS
COLON
CECUM
RECTUM
ANUS

B. Abdominal Organs

VILLI

C. Cross-Section of Small Intestine

Locate the pharynx. Food leaves the pharynx via the **esophagus** on its way to the stomach. The opening into the esophagus is dorsal to the glottis. Insert a blunt probe into the opening and follow the esophagus posteriorly through the thoracic cavity. In the thorax, the esophagus lies just dorsal to the trachea. The esophagus penetrates the diaphragm and continues posteriorly to join the bag-like organ, the **stomach**. The **cardiac sphincter** is the muscle which forms the boundary between the esophagus and the stomach. Refer to Figure 3 in identifying the rest of the digestive system.

The stomach can be divided into the larger anterior **cardiac** portion and the lower tapering **pyloric** portion. This posterior segment of the stomach joins the small intestine. Open one side of the stomach and examine its interior surface. Does it appear smooth or rough? Locate the **liver**, the large lobed, reddish-brown organ which lies posterior to the diaphragm. Notice that the liver consists of several lobes which are attached only at the dorsal and anterior margins. Among other functions, the liver produces bile, which emulsifies fats in the small intestine. The bile is temporarily stored in a small sac, the **gall bladder**. This organ may be found by lifting up the extreme right lobe of the liver to which the gall bladder is attached. At the posterior end of the stomach, locate the hard ring of smooth muscle, the **pyloric sphincter**. This sphincter muscle forms the boundary between the stomach and the **small intestine**. Constriction of this muscle prevents food from escaping into the **small intestine** before the stomach has finished processing it.

The portion of the small intestine into which the stomach empties is the **duodenum**. Locate a long, whitish, cauliflower-like organ, the **pancreas**, lying dorsal to the duodenum and the stomach. The digestive enzymes produced by the pancreas pass into the duodenum. Try to locate the **pancreatic duct**. It may be difficult to find. Also locate the **bile duct** as it enters the duodenum from the gall bladder.

To the left of the stomach observe the **spleen**, a reddish brown tongue-shaped organ which functions in destroying old red blood cells in the adult. Notice that the small intestine is coiled, therefore providing an increased area for digestion and absorption of food. The small intestine is also held in place by a mass of sheetlike membranes, the **mesenteries**. Slit open a short portion of the small intestine and find the **villi** with a stereomicroscope. The villi are microscopic finger-like projections which serve to increase surface area also.

The small intestine continues posteriorly, merging with the first of three segments of the large intestine, the **colon**, which is a compact, rounded mass of intestine bound firmly by mesentery. Find junction of the small and large intestines. The pig does not have an appendix. In man, the appendix is attached to the posterior end of the cecum. The straight most posterior section of the large intestine is the **rectum** which opens to the outside by way of the **anus**. The cecum, colon, and rectum make up the large intestine.

If time allows, sever the coiled small intestines just below the duodenum, and sever the colon at the point where it joins the rectum. Carefully cut the mesenteries holding the long intestinal sections in place so that it can be laid out in a straight line. How long are the intestines?

Locate as many of the structures as possible in the digestive system on the human torso model and color plates in the manual.

Circulatory System

The major **arteries** and **veins** are injected with latex. The arteries, which carry blood away from the heart, are injected with red latex, and the veins, which carry blood toward the heart, are injected with blue. Sometimes the pressure of injection causes the latex to cross capillary beds in some places. Therefore, veins occasionally contain some red latex, while arteries may have blue latex.

Arterial System (Refer to Figure 4, 6, 7)

Locate the **heart** in the thoracic cavity and carefully trim away the **pericardial sac** which surrounds the heart. The major portion of the heart is composed of the **right** and **left ventricles**. These are separated from one another on the ventral surface by the **coronary artery**. The left ventricle alone makes up the posterior tip, or **apex**, of the heart. Anteriorly and laterally are two dark projections, the right and left **auricles,** which are parts of the upper heart chambers, the **atria**. The term "auricle" comes from a root meaning "ear". The projections looked like ear flaps on each side of the heart. The auricles do not have any particular function.

FIGURE 4

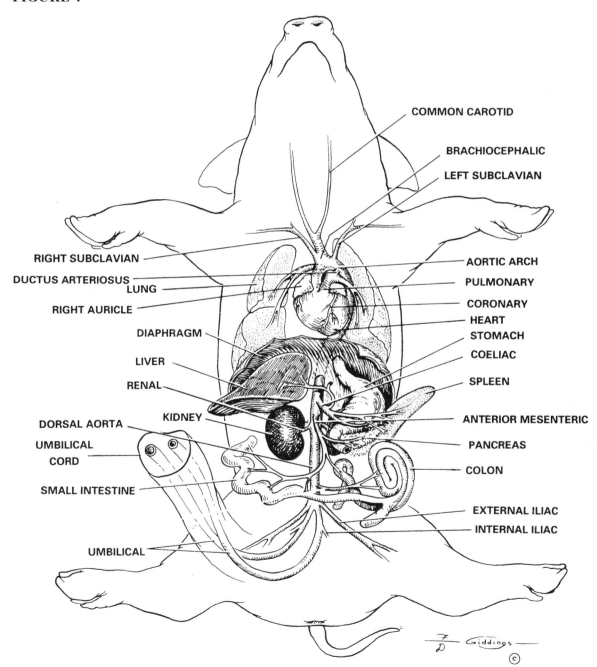

Locate the **pulmonary trunk**. It is a large vessel which carries blood from the right ventricle of the heart. This large artery crosses from the upper right over the ventral surface of the heart, and arches to over the dorsal side of the heart, where it branches. Carefully remove the connective tissue surrounding the pulmonary trunk and trace it until it branches by displacing the heart to the right (the pig's right) side. The **right pulmonary artery** goes to the right lung and the **left pulmonary artery** goes to the left lung or **aorta**, which carries blood from the left ventricle. It passes anteriorly for a short distance and makes a sweeping 180° left turn and comes to lie posteriorly along the dorsal wall of the thoracic and abdominal cavities.

FIGURE 5 The Fetal Pig (Major Veins)

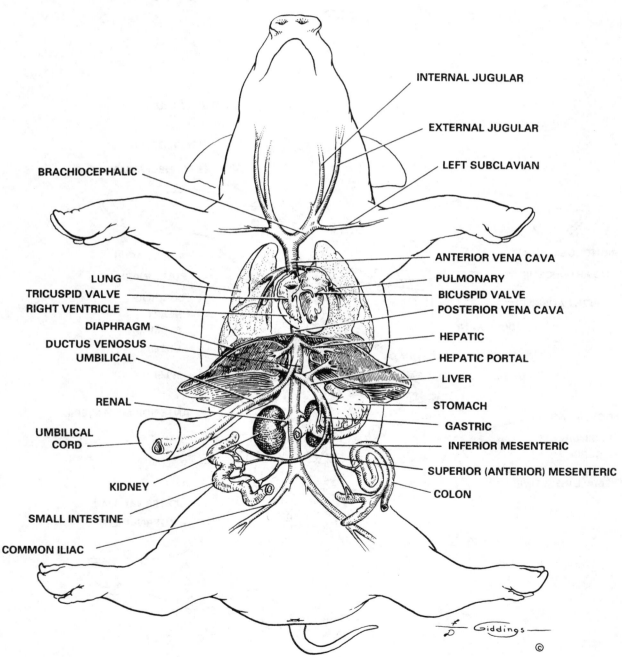

Observe that the pulmonary trunk connects directly to the aorta on the left side of the heart. After the trunk branches to the right and left lungs, the short, thick interconnecting vessel is known as the **ductus arteriosus**. The ductus arteriosus shunts blood into the **dorsal aorta**, bypassing the lungs. The lungs are nonfunctional in the fetal pig. After birth, the arterial duct normally closes off with connective tissue.

Carefully dissect away the connective and muscle tissues in the neck to expose the arterial branches going to the head.

Two major arterial branches emerge from the aortic arch a short distance from the heart. The first is the **innominate artery** or **brachiocephalic artery** which supplies blood to the right forelimb and head. This branch continues toward the head, then divides into two branches. The first is the **right subclavian artery** which supplies blood to the right forelimb and right ventral chest wall. The second branch continues anteriorly, dividing into two **common carotid arteries**, which supply blood to the head. The second branch off of the aortic arch is the **left subclavian artery** which carries blood to the left forelimb and the left ventral chest wall.

Find the dorsal aorta in the abdominal cavity. The first major branch is the **coeliac artery** which supplies the stomach, spleen and liver. The **anterior mesenteric artery** branches from the dorsal aorta just posterior to the coeliac artery. It supplies blood for the pancreas and the small intestine. Continue tracing the aorta posteriorly and locate the **renal arteries**, one leading to each kidney. Posterior to the renal arteries locate the paired **genital arteries** which are thread-like in size and lie on the extreme ventral surface of the dorsal aorta. The genital arteries lead and supply blood to the sex organs. At the posterior end of the body cavity, the dorsal aorta divides into two pairs of arteries that supply blood to the legs, the **external iliac arteries** and the **internal iliac arteries**. The two large **umbilical arteries** pass ventrally from the internal iliac arteries.

Venous System (Refer to Figure 5)

Find the large vein which enters the right atrium anteriorly. This is the **anterior vena cava** or **precaval vein**. There are four major veins which drain the anterior region of the body that unite to form this blood vessel. The two **external jugular veins** lie parallel to the carotid arteries. These are paralleled by two **internal jugular veins**. Also locate the **left** and **right subclavian veins** which drain each of the forelimbs, respectively.

Locate the **azygos vein** which lies to the left of and parallel to the dorsal aorta in the chest cavity. The azygos vein drains the muscles between the ribs and empties directly into the right atrium.

Locate the large **posterior vena cava** or **postcaval vein** which runs parallel to the dorsal aorta in the lower abdomen. Note where the two **renal veins** (one from each kidney) join the postcava at the kidneys. Posteriorly, the posterior vena cava will divide into the paired **common iliac veins**. Further subdivision of the common iliac veins does occur, but the veins are difficult to find.

Hepatic Portal Division of the Venous System (Refer to Figure 5)

When you originally opened the abdomen, you cut the **umbilical vein** which leads from the umbilical cord into the liver. One of its branches leads into the hepatic portal system, and the other branch, the **ductus venosus**, leads directly into the posterior vena cava. The ductus venosus allows some blood rich in oxygen and nutrients to be pumped out to the body without passing through a capillary bed in the liver. Try to find the **ductus venosus** which may have been cut when you opened up the pig.

The organs of the gastrointestinal tract are drained by the **hepatic portal system**. This is a set of veins which collects blood from the digestive tract and filters it through the liver before the blood enters the heart.

In order to expose the hepatic portal vessels move the stomach, spleen, pancreas, and small and large intestines to the left. They may not be injected with blue latex in your pig; therefore, only the largest vessels can be identified.

Locate the **hepatic portal vein**, the main vein of the hepatic portal system. It carries blood from the intestine to the liver. In the liver the blood passes through a capillary bed, where toxic materials are removed from the blood and the nutrient content of the blood is regulated by cells of the liver. Blood leaving this capillary bed eventually enters the posterior vena cava via the **hepatic veins**.

Locate the **superior (anterior) mesenteric vein** which is the union of branches from the many coils of the small intestine.

Also locate the **gastric vein** which drains the pyloric region of the stomach before it joins the hepatic portal vein.

The Heart

Do not remove the pig's heart. For this dissection, you will use the sheep's heart.

Open the sheep's heart by making a midventral slit if this has not already been done. Use Figure 7 as a guide. Blood leaving the right atrium passes through a one-way valve, the **tricuspid valve**, into the right ventricle. Find the **bicuspid**, or **mitral valve**, in the same position in the left side of the heart. The tough cords holding the edges of the valves in place are called **chordae tendinae**. The chordae tendinae, which prevent the valves from flapping up into the atria when the ventricles contract, are attached posteriorly to **papillary muscles**, columns of muscle arising from the wall of the ventricle. In the fetal pig, the **foramen ovale**, an opening between the two atria, allows the lungs to be bypassed. It is reduced to a closed depression at birth. Locate this depression in the interatrial septum. Look down into the stub of the aortic arch and pulmonary trunk and observe the flaps that compose the **semilunar valves**. These valves prevent backflow of blood from the arteries.

After finding the bold face parts given previously on the sheep heart, locate the same parts on the beef heart on display and the human torso model.

Urinary System (Refer to Figure 8)

Locate the paired **kidneys**, lying dorsally against the abdominal wall. Notice that each kidney is covered by a thin membrane, the **peritoneum**, on its ventral surface. Next, find the **adrenal gland**, a narrow whitish body about half an inch long which lies medially along the anterior edge of the kidney. Remove the peritoneum from a kidney. On the medial side of the kidney is a concave depression, the **hilum**. At this point the **renal artery** and **renal vein** attach, carrying blood into and out of the kidney, respectively. The hilum is also the point where the **ureter** leaves the kidney, carrying urine to the **urinary bladder**. Expose one of the ureters by picking away the peritoneum covering it.

FIGURE 6 Sheep Heart Lateral

FIGURE 7 Sheep Heart (L.S.)

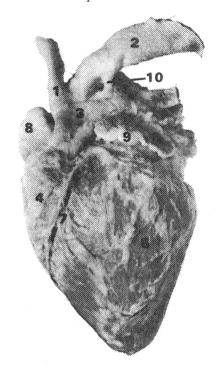

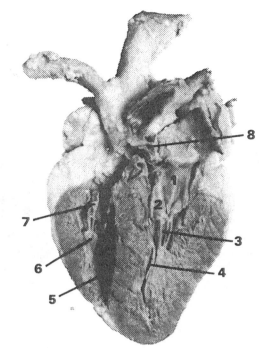

1. BRACHIOCEPHALIC
   ARTERY
2. AORTA
3. PULMONARY ARTERY
4. RIGHT VENTRICLE
5. LEFT VENTRICLE

6. APEX
7. ANTERIOR
   LONGITUDINAL SCILCUS
8. RIGHT AURICLE
9. LEFT AURICLE
10. DUCTUS ARTERIOSUS

1. LEFT ATRIUM
2. BICUSPID VALVE
3. CHORDAE TENDINAE
4. LEFT VENTRICLE

5. RIGHT VENTRICLE
6. PAPILLARY MUSCLE
7. TRICUSPID VALVE
8. SEMILUNAR VALVE

During fetal life, urine exits from the anterior end of the urinary bladder by way of the **allantoic duct**, through the umbilical cord to the placenta. In the adult male, urine is voided to the outside of the body by the **urethra**. This is a tube-like structure which passes posteriorly from the urinary bladder for a short distance and then turns anteriorly and ventrally to enter the **penis**. In the adult female the urethra continues posteriorly to enter the **urogenital sinus**, or **vestibule**, which lies approximately one-half of an inch from the urogenital opening.

Using your razor blade or scalpel, section one of the kidneys in place, cutting from the lateral or the medial side on a plane parallel to the dorsal side of the animal. Note that at the center of the medial portion of the kidney is an irregular cavity, the **pelvis**. Here the urine is collected and through this region the branching blood vessels pass. The pelvis is continuous with the ureter.

The outermost portion of the kidney, the **cortex**, shows many small striations perpendicular to the outer surface. This region and the **medulla** region, which lies medially to it, are composed of thousands of minute excretory tubules called **nephrons** associated intimately with capillaries. Urine is formed in these regions and then drains to the pelvis.

FIGURE 8 The Fetal Pig (Urogenital Organs)

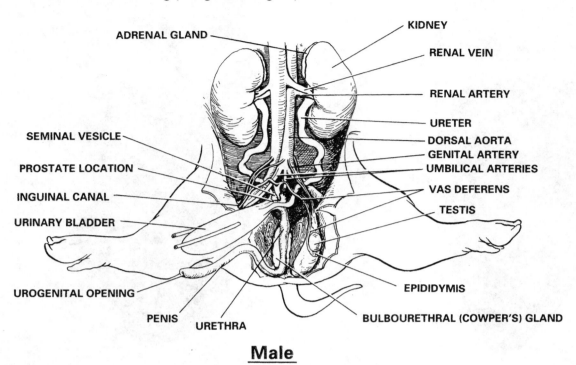

Male

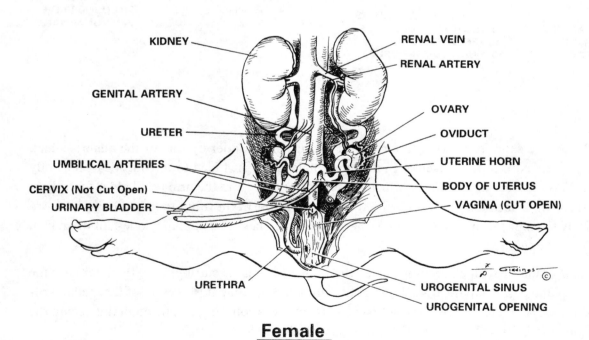

Female

Reproductive System

The reproductive system in mammals is closely associated with the urinary system. Together, these form the **urogenital system**. In this exercise, they will be discussed separately, but in Figure 8 they are combined.

Female Reproductive System

Pull the umbilical cord posteriorly and find the urinary bladder. The urinary bladder continues posteriorly joining the urogenital sinus via the urethra. Cut along one side of the urogenital sinus and through the cartilage of the pelvic girdle. Now you should be able to lay the legs out flat.

Referring to Figure 8, locate the pair of gonads the **ovaries**, about one half an inch posterior to each kidney. The ovary is a small, bean-shaped, light colored structure. The lateral and dorsal surface of each ovary is partially covered by the mouth of the **oviducts**, or **fallopian tubes**. The oviducts lead posteriorly to the **horns of the uterus**. The horns join along the midline to form the **body of the uterus**. Posterior to the body of the uterus, locate the **vagina**. The vagina joins with the **urethra** forming the **urogenital sinus**, or **vestibule**, an area shared by the urinary and reproductive systems. The **glans clitoris** a small rounded papilla arises from the ventral floor of the urogenital sinus. This structure may not be apparent in your specimen. The **glans clitoris** clitoris is the female homolog to the penis in the male. The urogenital sinus opens to the outside via the **urogenital opening**, just ventral to the anus. Now locate the dorsally directed **genital papilla**.

Male Reproductive System

Referring to Figure 8, locate the pair of gonads, the **testes**, where sperm is produced. The testes begin their development in the body cavity, just posterior to the kidneys. Before birth they descend into the paired **scrotal sacs**, located between the hind legs and just ventral to the tail. In younger specimens, the testes have not yet descended and are located between the abdominal cavity and the scrotal sacs. Each scrotal sac is connected to the body cavity by the **inguinal canal**.

Locate the opening of the left inguinal canal. Then make an incision through the skin and muscle layers from a point over this opening to the left scrotal sac. In this way, the canal and whole sac will be exposed. Open the sac and find the testis. Note the much-coiled tubule, the **epididymis**, which lies along the surface of the testis. This is continuous with the sperm duct, or **vas deferens** which passes back toward the body cavity. In the body cavity, the sperm ducts from each testis loop over the umbilical arteries and ureters and unite dorsally at the posterior end of the urinary bladder with the **urethra**.

Near where the vasa deferentia (singular: vas deferens) join the urethra and dorsal to the urethra, locate a small pair of light colored glands, the **seminal vesicles**. A small rounded gland, the **prostate**, may be found in the dorsal surface of the urethra just posterior to the junction of the urinary bladder and the urethra. The prostate may be difficult to locate.

Cut through the cartilage of the pelvic girdle on the left side. Trace the urethra posteriorly until it makes a U-turn and proceeds anteriorly until it opens to the outside via the urogenital opening of the penis. Now locate a pair of white glands, the **bulbourethral (Cowper's) glands**, one on each side of the urethra near the U-turn. These three types of glands form a fluid, **seminal fluid**, which together with sperm constitutes the **semen**. Seminal fluid nourishes and provides transport for sperm from the epididymis.

FIGURE 9 (*a*) Lateral view of fetal pig musculature. (*b*) Ventral view of fetal pig musculature.

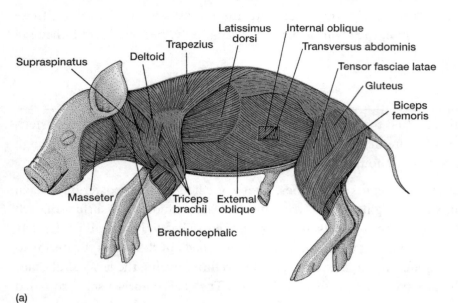

(a)

(b)

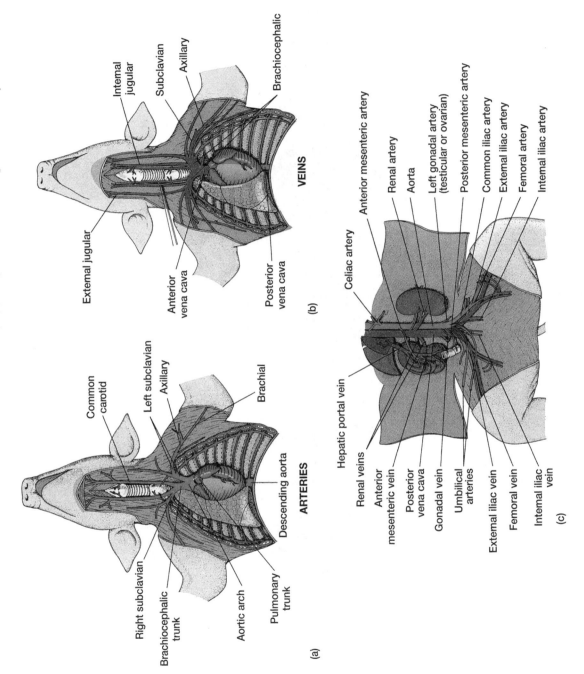

FIGURE 10 (*a*) arteries and (*b*) veins of the anterior body. (*c*) Major vessels of the posterior body.

FIGURE 11 Major visceral organs of a fetal pig.

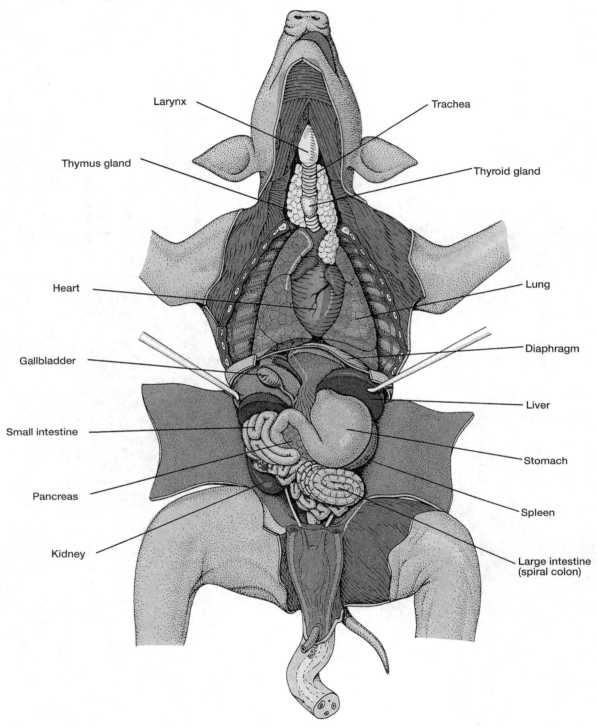

FIGURE 12 Urinary and reproductive organs of (a) a female and (b) a male fetal pig.

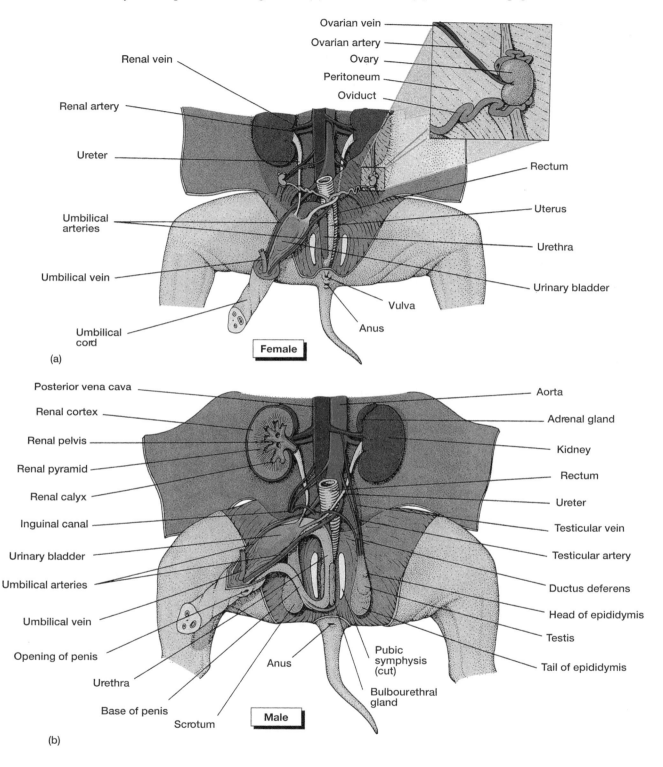

Conversion of Metric Units to English Units

UNITS OF LENGTH

The meter (m) is the basic unit of length.

1 m = 39.4 inches (in)
    = 1.1 yard (yd)   1 in = 2.54 cm
1 km = 1000 m = 10^3 m
    = 0.62 miles (mi)   1 ft = 30.5 cm
1 cm = 0.01 m = 10^{-2} m
    = 0.39 in = 10 mm   1 yd = 0.91 m
1 nm = 10–9 m = 10^{-6} mm
    = 10 angstroms (Å)   1 mi = 1.61 km

Units of area are squared (two-dimensional) units of length.

1 m^2 = 1.20 yd^2 = 1550 in^2 = 1.550×10^3 in^2

1 hectare = 10,000 square meters (m^2) = 2.47 acres

Measurements of area and volume can use the same units.

1 m^3 = 35.314 ft^3 = 1.31 yd^3
1 cm^3 (cc) = 0.000001 m^3 = 0.061 in^3

Units of Mass

The gram (g) is the basic unit of mass.
1 g = mass of 1 cm^3 of water at 4°C = 0.035 oz
1 kg = 1000 g = 10^3 g = 2.2 lb

Units of Volume

The liter (L) is the basic unit of volume. Units of volume are cubed (three-dimensional) units of length.

1 liter = 1000 cm^3

1 liter = 2.1 pints = 1.06 qt   1 cup = 240 mL

1 liter = 0.26 gal = 1 dm^3

1 mL = 0.035 fl oz

Units of Temperature

5 × degrees Fahrenheit = (9 × degrees Celcius) + 160

For example:

40°C = 104°F (a hot summer day)

75°C = 167°F (hot coffee)

−5°C = 23°F (coldest area of freezer)

37°C = 98.6°F (human body temperature)

BIOLOGICAL TERMS

Learn the meaning and application of the following terms as soon as possible. These terms are used in describing the structures of animals and give direction for study and dissection.

aboral—pertaining to the region most distant to the mouth

anterior—the front, head end, or forward moving end of a bilaterally symmetrical animal

asymmetrical—an organism, body or body part that cannot be divided into two or more equivalent parts

bilateral symmetry—an organism, body or body part that can be divided into two equivalent parts, left and right halves, each a mirror image of the other

caudal—pertaining to or toward the posterior part or tail of an organism or body

cephalic—pertaining to or toward the anterior part or head of an organism or body

cross section—a section perpendicular to the long axis of the body

distal—away from the central part of the body or place of attachment

dorsal—pertaining to or toward the back or upper surface

lateral—pertaining to or toward one side

longitudinal—lengthwise or pertaining to the long axis of the body

median—pertaining to, near, or toward the middle line of the body

metamerism—body, externally, internally, or both, which is composed of a number of similar and homologous parts (segments, somites, or metameres) as seen in higher animal forms

oral—pertaining to the mouth

posterior—the hind part; toward the tail end of a bilateral symmetrical animal

proximal—toward or near the central part of the body

radial symmetry—an organism, body or body part having similar equivalent parts arranged around a common axis

sagittal plane—a plane that divides the body into symmetrical right and left halves

transverse plane—any plane at right angles to a sagittal plane; a cross section

ventral—pertaining to or toward the belly or lower side

SOME COMMON GREEK AND LATIN PREFIXES, SUFFIXES, AND ROOTS WITH THEIR BASIC MEANINGS

a-; an-: not; without; an-emia; a-cephal-ous; a-poda

ad-; ac-; af-; ag-; ap-: to; toward; ad-hesion; af-ferent; ag-glutin-ate; ad-ductor; ap-pendage

ana-: up; ana-bol-ism; ana-phase; ana-tomy; an-oxia; an-ion

ant-; anti-: against; opposite; anti-biotic; anti-gen; anti-bodies

aqu-; aqua-; aquae-: water; liquid; aquat-ic; aqu-arium

-arium; -ary: place where something is kept, produced, or studied; api-ary; avi-ary; aqu-arium

arthr-; arthro-: joint (of body); arthr-itis; arthro-pod

auto-; aut-: self; spontaneous; auto-nomy; auto-nomic; auto-tomy; auto-troph

bi-; bin-: two; twice; double; bi-cuspid; bin-ocular

bio-:life; living; bio-logy; biot-ic; bio-genesis; bio-me; sym-bio-sis

blast-; blasto-: bud; cell; blast-ula; ecto-blast; osteo-blast; triplo-blastic

card-; cardi-: heart; cardi-ac; peri-card-ium

carn-; carno-; carni-: meat; flesh; carni-vore; carn-al

cephal-; cephalo-: head; cephalo-pod; a-cephal-ous; cephalo-thorax

chrom-; chromo-; chromat-: color; hue; chromo-some; chromat-id; chromat-in; chromo-nema

circum-: around; surrounding; circumpharyngeal

cyst-; cysti-; cysto: bladder; bag; sac; cyst; chole-cyst; sporo-cyst; tricho-cyst

cyt-; cyto-: cell; cavity; cyto-logy; cyto-kinesis; cyto-plasm; leuco-cyte; phago-cyte; choano-cyte

derm-; dermo-; dermat-: skin; hide; epi-dermal; derm-al; ecto-derm

di-; dis-; dipli-; diplo: two; twice; double; di-ptera; dipl-oid; di-morphic; dio-ecious; diplo-blastic; di-hybrid

di-; dia-; dis-: across; through; dissect; dia-phragm; di-gestion

e-; ef-; ex-: from; out of; without; ef-ferent; e-dentate; exo-skeleton; ef-fector; e-gest; ex-cretion

ec-; ecto-: from; outside; without; ecto-derm; ecto-zoa; ecto-plasm

end-; endo-; ent-; ento-: within; inside; endo-derm; endo-cardium; endo-crine; endo-genous; endo-skeleton; endo-thelium

ep-; epi-: on; upon; over; epi-cardium; epi-dermal; epi-glottis; epi-dermis; epi-thelium

-fer; -freous: bearing; producing; ovi-ferous; omni-ferous; roti-fer; pori-fer-a

gam-; gamo-; -gamy: marriage; fusion; syn-gamy; iso-gamy; gam-ete; auto-gamy

gastro-; gast-: stomach; gastr-ic; gastro-pod; gastro-coel; gastro-dermis; gastr-ula

gen-: be born; producing; gen-es; genet-ics; anti-gen; geno-type; meta-gene-sis; oo-gene-sis; partheno-gene-sis

gloss; -glott: tongue; language; hypo-glossal; glott-is; epi-glottis

hem-; hemo-; hemat-: blood; hem-al; hemato-gram; hemo-rrhage; hemo-globin; hemo-coel

hepat-; hepato-: liver; hepat-ic; hepat-itis; heap-rin

heter-; hetero-: other; different; hetero-genous; hetero-troph

homeo-; homo-: like; similar; same; homo-genous; homo-logue

hyper-: above; beyond; hyper-tonic; hypo-glossa; hyper-emia; hyper-dermic; hyper-trophy

hypo-: under; loss; below; hypo-tonic, hypo-glossal

is-; iso-: equal; similar; iso-tonic; iso-gamy

-log; -logy: word; discourse; study; bio-logy; zoo-logy; embryo-logy

mamm-; mamma-; mammae-: breast; nipple; mamm-al; mamm-ary

mes-; meso-: half; middle; mes-entery; meso-derm; mes-encephalon; meso-nephros

micro-; micr-: small; little; micro-scope; micro-be; micro-biology; micr-on

mon-; mono-: one; single; mono-saccharide; mono-cular; mono-ecious; mono-hybrid

nephr-; nephro-: kidney; nephr-idia; nephro-stome; nephr-it is; pro-nephros; nephr-on

nucle (nux): nut; nucle-us; nucleo-plasm; nucleo-tide

ocul-; oculus-oculi: eye; graft; ocul-ar; bin-ocul-ar; ocell-us; in-ocul-ation

-oid: like; in the form of; amoeb-oid; coll-oid; dipl-oid

omni-: all; omni-vorous

oss-; os-; ossi-; oste-; osteo-: bone; oss-eous; ossi-fy; osteo-blast; per-oste-um; ossi-cle

ov-; ovum; ovi-: egg; ov-ary; ovi-duct; ovi-parous; ovo-viviparous

-parous: producing; ovi-parous; vivi-parous

pod-; podi-; podo-: foot; pseudo-pod; rhizo-pod; cephalo-pod; tetra-pod

poly-: many; much; poly-morph; poly-saccharide

pro-; proto-: first; pro-stomium; pro-tists; proto-plasm

pseudo-; pseudo-: false; sham; pseudo-pod; pseudo-morph; pseudo-coel

sec-; sect-; seg-: cut; sec-tion; bi-sect; in-sect; vivi-sect; seg-ment

semi-: half; semi-permeable

soma-: body; chromo-some; auto-some

sub-: under; below; sub-esophageal; sub-lingual; sub-maxillary; sub-cutaneous

super-; sur-; supra-: over; beyond; more; supra-esophageal

ventr-; venter-; ventri-: stomach; ventr-al; ventri-cle

vivi-: live; ovi-vivi-parous; vivi-sect

-vore; -vorous: eating; devouring; carni-vore; herbi-core; omni-vore; insecti-vorous

zo-; zoa-; zoo-: animal; zoo-logy; proto-zoa; zoo; holo-zo-ic; metro-zo-an; spermato-zoa

zyg-; zygo-: pair; couple; yoke; zygo-te; zygo-spore; hetero-zyg-ous; homo-zyg-ous